Own The Insight

Turn First-Party Data Into Revenue Faster Than Your Competition Can React

Lisa L. Fagen

Fagen Omni Enterprises, Inc.

ISBNs:
ISBN 979-8-9930038-2-5 (hardback)
ISBN 979-8-9930038-1-8 (paperback)
ISBN 979-8-9930038-0-1 (eBook)

Disclaimer: All real companies are referenced with permission. “Northbridge Brands” is a fictional manufacturer used for illustrative purposes. Promolytics (Promolytics.net) is a real platform mentioned as an example of promotion insight technology.

Business Practices and Terminology Disclaimer: This book explores established business concepts, methodologies, and industry-standard terminology commonly used in marketing, data analytics, and business strategy. Terms such as "first-party data," "third-party data," "customer lifetime value," "progressive profiling," "data maturity models," "KPIs," "ROI," and similar business vocabulary are commonly used in the industry. They are understood in their generally accepted meanings.

The business practices, strategies, and methodologies discussed herein, including but not limited to customer segmentation, A/B testing, data governance, consent management, and marketing attribution, represent widely adopted industry standards and best practices. These concepts have evolved through collective industry knowledge and are not proprietary to this work.

While this book presents these concepts through the unique lens of the "Own the Insight" methodology and the fictional Northbridge Brands case study, readers should understand that many underlying principles are established business practices. The author's contribution lies in the specific synthesis, application, and presentation of these concepts within the INSIGHT Framework and the Consumer Insights Flywheel approach.

COPYRIGHT

Any resemblance to specific company practices or proprietary methodologies is coincidental, as these represent common industry approaches.

Framework and Model Attributions

The Flywheel Concept: The business flywheel metaphor, popularized by Jim Collins in his book "Good to Great" (2001), has been notably adopted by companies such as Amazon. Our Consumer Insights Flywheel adapts this concept specifically for activating first-party data.

Data Maturity Models: The five-stage maturity model (Blind → Reactive → Aware → Strategic → Intelligent) builds upon established data maturity frameworks, including:

1. The Capability Maturity Model (CMM) was initially developed at Carnegie Mellon University
2. Gartner's Analytics Maturity Model
3. TDWI's Analytics Maturity Model. Our adaptation focuses specifically on first-party data capabilities.

Progressive Profiling: This technique has been a standard practice in marketing automation since the early 2000s, popularized by marketing automation platforms like Marketo, HubSpot, and Pardot.

INSIGHT Framework: While the INSIGHT acronym and specific seven-stage process are original to this work, it incorporates established practices:

- Data normalization and identity resolution (standard in Customer Data Platforms)
- Segmentation (foundational marketing practice dating to the 1950s)
- Predictive analytics (an established field of data science)
- Marketing automation (industry practice since the 1990s)

Build-Measure-Learn: Our iterative approach draws inspiration from Eric Ries's "Lean Startup" methodology (2011), adapted for marketing data applications.

Value Exchange Principle: The concept of value exchange in data collection has been discussed extensively in privacy and marketing literature, including works by the Direct Marketing Association and various privacy researchers.

Data Governance Frameworks: Our governance approach incorporates elements from:

- ISO/IEC 38500 (IT Governance)
- COBIT framework
- Generally Accepted Privacy Principles (GAPP)

Visual Models:

- Figure 1 (Consumer Insights Flywheel): Original visualization inspired by Jim Collins' flywheel concept
- Figure 2 (INSIGHT Framework): Original framework and visualization
- Figure 3 (First-Party Data Maturity Model): Original adaptation of established maturity model concepts

The author gratefully acknowledges these foundational concepts and the broader community of practitioners who have developed and refined these methodologies over time.

First edition 2025

Dedication:

To my family and friends, thank you for your unwavering support and endless encouragement.

Acknowledgements:

Thank you to my colleagues, early readers, and customers whose questions helped shape these pages.

Table of Contents

Prologue

10 Ways First-Party Data Changes the Game

While your competitors debate market trends in conference rooms, your customers are voting with their wallets in real time. Every scan, click, redemption, and conversation is a signal that reveals exactly what to make, where to sell, and how to grow, if you know how to capture and connect those signals to actual purchases. This book shows how to capture those signals with permission, connect them to purchases, and act while it still matters. You will learn a simple system for collecting, unifying, and activating first-party data so every campaign becomes smarter than the one before. Most importantly, you will know exactly what your customers think about your product, what matters most to them, and what new ideas excite them now, rather than relying on a generic industry persona. The goal is a practical advantage that compounds, not theory.

The data is clear: companies that master first-party data strategies significantly outperform their competitors. Forrester's 2024 research indicates that incorporating first-party behavioral data into marketing strategies can improve ROI by 72% and reduce customer acquisition costs by 83%.[1] Meanwhile, 83% of companies that exceeded their revenue goals used first-party data to inform their

1. First-Party Data Statistics Every Marketer Should Know. Aishika Das. Last updated Jan 2, 2025

strategies.[2] and digitally mature brands utilizing first-party data in advanced marketing see up to 2.9x higher revenue uplifts. [3] Companies still relying on generic insights and third-party data, watch competitors capture market share with precision they can't match.

Here's how first-party data transforms every aspect of your business:

Revenue Growth

1. **Turn tastings into traceable relationships by capturing** permissioned first-party signals at the moment of sampling. Next, link these signals to offers and redemptions, and verify purchases with receipts. **See which venues, messages, and partners move customers from their first sip to repeat purchases,** and then optimize the next activation in real-time.

2. **Discover which promotions drive purchases.** Replace gut feel with proof by tying the offer to verified transactions so you scale only the winners.

3. **Score influencers and ads by revenue, not vanity metrics.** Attribute sales to creators, placements, and messages so you fund what converts.

4. **Optimize value exchange in real time.** Match offers to individual willingness-to-pay signals, ensuring every customer receives the right incentive at the right moment to maximize both conversion and margin.

---

2. PGM. The Power of First-Party Data. Jul 27, 2025

3. AVAUS. First-party data benchmarks – what results can you expect?

Customer Intelligence

1. **Decode preferences and dislikes at a granular level.** Learn flavors, formats, price points, and use cases so your targeting feels personal and relevant.

2. **Fuel product innovation with real requests.** Turn open-ended feedback and purchase patterns into a roadmap that customers are already waiting to buy.

3. **Eliminate revenue leaks in the journey.** Identify where prospects drop off and address packaging, pricing, or messaging issues that block conversions.

4. **Detect loyalty in real time.** Notice repeat purchase cadence and reward high-value customers before they drift to alternatives.

Market Power

1. **Transform retailer relationships from vendor to partner.** Present verified consumer demand data to secure premium shelf placement, negotiate better terms, and co-create promotional strategies that drive mutual profit growth.

2. **Ultimately, create a compounding asset.** Every signal enriches your database, making each campaign smarter and each dollar more efficient over time.

By the end of this book, you will stop renting data and start owning the relationship. Replace rented lists with first-party signals. Swap static personas for real behavior and guesswork for verified results. Invest only in what performs, allocate budget to winning channels, eliminate waste, and negotiate with proof of demand. The playbooks ahead are transparent, ethical, and actionable. The window for building this advantage is closing. As privacy regulations tighten and third-party data becomes scarce, first-party relationships emerge as the only sustainable competitive moat.

Introduction:

The Great Data Awakening

Every executive faces the same frustrating paradox: more marketing data than ever before, yet less certainty about what actually drives results. Teams present impressive dashboards showing engagement spikes, click-through improvements, and social media wins while customer acquisition costs climb and revenue growth stagnates. The disconnect isn't a data problem; it's an insight ownership problem.

Third-party data tells you what happened to someone else's customers. First-party data reveals what your customers actually want, when they want it, and how much they're willing to pay. The difference between renting insights and owning them determines whether you're reacting to market trends or creating them.

The Northbridge Journey: How One Company Turned Customer Signals Into Competitive Advantage

Picture the opening of a quarterly review meeting at a midsize manufacturer. Dashboards glow with promising numbers: social engagement is up, email opens are at an all-time high, and a recent promotion drove thousands of coupon scans. Yet when the CFO reviews the revenue sheet, a critical discrepancy becomes apparent: the ROI per promotion is alarmingly low, highlighting missed opportunities and stagnant growth. The optimism shifts to unease as no one can pinpoint which spending created true value. By leaning on outside data and chasing vanity metrics, the company overlooks the insights that drive growth.

Owning the insight changes everything. It's not just about data; it shapes your entire relationship with your customer. When Northbridge owns the insight, it means connecting understanding to action. Imagine a busy parent who now receives truly useful product suggestions, or a passionate foodie discovering fresh inspiration. First-party data, when turned into actionable insight, connects marketing, sales, leadership, strategy, and product development around customer behavior. This is how you unlock value and enrich lives.

To bring this transformation to life, this book guides readers through the challenges and successes of a fictional company, "Northbridge Brands," as it moves from blind spend to intelligent execution. Along the way, we will demonstrate how to:

1. **Diagnose hidden costs** created by untracked promotions and rented audiences.

2. **Establish a first-party foundation** built on consent, data quality, and lawful capture.

3. **Design collection around building stronger customer relationships**, creating permission-based moments that add value for the customer.

4. V**alidate digital performance** with first-party proof, creating a complete view of customer journey effectiveness.

5. **Create a Consumer Insights Flywheel** that accelerates learning and fosters a deeper understanding of the consumer's needs and preferences.

6. **Operationalize the INSIGHT Framework** to turn signals into action across campaigns, channels, and offers.

7. **Create a cross-functional insights team** to activate findings across

Marketing, Product, Sales, and Operations.

8. **Scale innovation with smart guardrails** that accelerate experimentation while ensuring consistent growth across all departments.

The result is practical clarity for every leader: Product Development sees direct customer requests and plans the next release with confidence; Marketing funds the channels that win; Operations balances inventory by retailer and region to match demand; Sales negotiates from strength with source-level proof of demand; and the C-suite sees a real-time view of demand and performance.

To support your journey, most chapters conclude with an Executive Reflection Prompt and a Checklist, enabling you to translate ideas into immediate progress. Use them. Insight is not a one-time project. It is a habit that compounds.

Regardless of your role, whether you lead Marketing, work in Sales, Strategy, Finance, Product Development, or the C-suite, the principles remain the same. A unified source of real-time, first-party data aligns teams, eliminates silos, and accelerates decision-making. The opportunity lies in building this capability now, setting your organization up to move faster and smarter in a rapidly evolving market.

The companies that master this transformation in the next 18 months will own significant competitive advantages. Those that wait will find themselves playing catch-up in a market where customer relationships, not rented data, determine winners.

Begin building your competitive moat with Chapter 1.

Chapter One

Insight Blindness:

How Data Abundance Creates Strategic Confusion

Before you can build a first-party data strategy, you must diagnose what's broken in your current approach to customer understanding. This chapter introduces the concept of 'insight blindness', the costly condition where organizations collect abundant data but fail to extract actionable intelligence that drives growth.

We'll examine the six warning signs that reveal insight blindness in action, quantify the hidden costs of operating without a clear understanding of customers, and identify the organizational habits that keep companies trapped in reactive decision-making. Most importantly, we'll establish why first-party data serves as the antidote to insight blindness and how unified customer intelligence transforms every business function from marketing to product development.

By the end of this chapter, you'll have a framework for auditing your organization's current insight capabilities and a clear roadmap for moving from data-rich but insight-poor operations to intelligence-driven competitive advantage.

When Data Abundance Becomes Strategic Blindness: The Northbridge Wake-Up Call

On a rainy Monday morning, Evelyn Harper, VP, Marketing Strategy, and Insights at Northbridge Brands, stared at a perplexing quarterly report. Sales had plateaued despite record social engagement and email performance. Marketing spend was increasing while returns were diminishing, raising concerns about the real impact of these metrics on revenue growth. When the CFO pressed for explanations, no one could connect their impressive dashboards to tangible business outcomes such as sales or profit.

Northbridge was experiencing insight blindness; the inability to transform abundant data into actionable understanding. Despite collecting customer information across multiple touchpoints, the company couldn't answer fundamental questions: Why had it's flagship product lost 15% market share in six months? Which marketing channels actually drove profitable customers? What did their best customers want next?

The wake-up call came when a smaller competitor captured significant market share using precise customer insights to launch targeted offerings. While Northbridge debated vanity metrics in conference rooms, data-driven competitors were stealing customers with surgical precision.

What Is Insight Blindness?

Insight blindness isn't a lack of data; it's the failure to convert customer signals into actionable insights and competitive intelligence. Organizations suffering from this condition often have sophisticated measurement systems but can't explain why their numbers change or predict what customers will do next.

Three Clear Warning Signs:

Unexplained Performance Shifts: Revenue or engagement numbers fluctuate dramatically, but teams can only speculate about causes. Northbridge watched a key customer segment defect to competitors without understanding whether price, product features, or service quality drove the exodus.

Decision-Making by Intuition: In the absence of customer intelligence, leaders default to experience and gut instinct. Northbridge's product roadmap relied heavily on executive opinions rather than evidence of what customers actually wanted.

Reactive Strategy: Companies consistently respond to market changes after competitors have already capitalized on opportunities. Northbridge discovered emerging customer preferences through industry reports months after agile competitors had launched solutions.

The Hidden Costs of Operating Blind

Insight blindness creates measurable business damage across every function:

Marketing Waste: Northbridge's marketing efficiency declined by 34% over 18 months as campaigns targeted broad audiences instead of high-value customer segments. Without understanding which channels drove profitable customers, the team increased spending on vanity metrics while acquisition costs continued to rise.

Innovation Failure: Three consecutive product launches failed, costing $2.1 million in development waste. Customer feedback data existed in service logs, but no system connected those signals to product development priorities. Competitors using systematic customer intelligence launched successful products while Northbridge built features nobody wanted.

Competitive Vulnerability: Customer lifetime value declined 22% while data-driven competitors grew theirs by capturing Northbridge's most profitable segments. Insight-blind companies can't defend against precision attacks because they don't understand what makes their customers loyal.

Why Smart Companies Go Blind

Three organizational habits create and perpetuate insight blindness:

Fragmented Data Architecture: Different departments collect customer information in isolation, resulting in a fragmented data architecture. Marketing tracks campaigns, sales logs interactions, and service records complaints, but no system unifies these signals. Northbridge had customer touchpoints across twelve different platforms with no integration, making comprehensive customer understanding impossible.

Limited Analytical Capability: Teams generate descriptive reports (what happened) but lack diagnostic or predictive capabilities (why it happened, what's next). Northbridge could provide you with last quarter's metrics, but couldn't explain the performance drivers or forecast changes in customer behavior.

Poor Information Flow: Even when insights emerge, they remain trapped within departments. Northbridge's customer service team identified emerging product complaints months before they impacted sales; however, this intelligence never reached the product development or marketing teams, who could have responded proactively.

The First-Party Data Solution

First-party data breaks the insight blindness cycle by providing relevant, real-time intelligence about your actual customers rather than generic market trends. Unlike third-party data that tells you what happened to someone else's customers, first-party signals reveal exactly how your customers behave, what they value, and where opportunities exist.

Unified Customer Intelligence: When properly collected and integrated, first-party data connects previously isolated touchpoints into a comprehensive view of customer journeys. Northbridge began seeing how social media interactions influenced in-store purchases, which email campaigns drove repeat buying, and what product features correlated with customer satisfaction.

Predictive Capability: Real-time first-party data enables proactive decision-making rather than reactive responses. Instead of relying on industry reports to discover customer trends, companies can identify shifts in their own customer base and respond promptly.

Competitive Differentiation: While competitors access the same third-party market research, your first-party data provides unique insights that create sustainable advantages. Northbridge's transformation began when leadership recognized its customer data as its most valuable strategic asset.

Executive Reflection

"The hardest part wasn't admitting we were blind; it was quantifying how much that blindness cost us. When we mapped our failed initiatives against available customer data, we identified $3.2 million in preventable losses over a two-year period. The breakthrough came when we stopped treating customer data as departmental property and started viewing it as enterprise intelligence. Now, every major decision begins with the same question: 'What do our customers' behaviors tell us about this choice?'" — Evelyn Harper, Director of Brand Strategy, Northbridge Brands

Chapter 1 Checklist: Diagnosing Insight Blindness

- **Audit Unexplained Trends:** List three recent performance changes your team couldn't fully explain. These gaps indicate where customer intelligence could provide answers.
- **Map Data Fragmentation:** Document where customer data lives across your organization. Identify silos preventing comprehensive customer understanding.
- **Assess Decision-Making Sources:** Review your last five strategic decisions. What percentage relied on first-party customer data versus intuition or third-party reports?
- **Calculate Blind Spot Costs:** Estimate losses from failed campaigns, product launches, or missed opportunities that customer intelligence could have prevented.
- **Evaluate Information Sharing:** Determine if customer insights dis-

covered by one department reach decision-makers in other functions within 30 days.

- **Secure Leadership Commitment:** Present insight blindness as a competitive risk requiring systematic first-party data strategy investment.

By completing the checklist above, your organization will have taken the first step: recognizing the extent of insight blindness in your current operations. In the following chapters, we will examine how to methodically build upon this foundation, progressing from merely recognizing the problem to systematically addressing it by owning the insight and transforming first-party data into your brand's most powerful asset.

Chapter Two

The Hidden Cost of Not Knowing:

When Unverified Spend Drains Budgets and Starves Intelligence

Most marketing budgets contain a hidden tax that finance teams rarely account for: the cost of unverified spending. This chapter reveals how promotional dollars are often diverted into unmeasured activities, resulting in both immediate waste and lost learning opportunities that accumulate over time.

We'll introduce the Hidden-Cost Ledger, a framework for quantifying four types of measurement blindness that drain budgets and starve organizations of customer intelligence. You'll learn to calculate your Unverified Spend Rate, measure Decision Lag Costs, and implement verification systems that turn every promotional dollar into both revenue and insight.

By the end of this chapter, you'll have concrete tools for eliminating waste while building the first-party data foundation that powers intelligent marketing decisions.

The Finance Room: Seeing the Blind Spot

When Northbridge's finance team audited last quarter's $800K promotional spend, they discovered $336K, 42% of the budget, had generated no verifiable customer outcomes. A $50K sampling tour generated social buzz but yielded zero tracked purchases. Print ads worth $75K reached audiences they couldn't identify. Event activations costing $120K collected no follow-up data. The company was paying twice: once for the promotion and again for the lost intelligence.

Why Hidden Costs Linger

That $336K figure represented more than accounting waste; it revealed a systemic problem. Northbridge wasn't just losing money on unverified promotions; they were trapped in a cycle that guaranteed continued waste. Without mechanisms to measure promotional effectiveness, the company continued to repeat expensive mistakes while missing optimization opportunities.

The hidden costs persist because most organizations operate with four fundamental measurement gaps:

1. **Untracked promotions**: Coupons redeemed without knowing which channel drove interest; events held with no post-event feedback. Dollars are spent, insight lost.

2. **Spray and pray buys**: Third-party segments appear large but often fail to align with actual customers.

3. **Lagging indicators**: Teams celebrate impressions and clicks that are not tied to sales, while revenue remains stagnant. Vendors' reports take months to deliver; by then, the data is stale and you're steering with yesterday's map.

4. **Opportunity cost**: Time and budget tied up in underperforming tactics leave no room to test new, potentially higher-return ideas.

Evidence sidebar: the industry context

Marketers report sharp momentum behind QR-enabled experiences and the broader transition to data-rich 2D codes, with a 93% increase in QR use over the past 12 months. Global industry bodies are targeting the wide adoption of GS1-standard QR codes at retail by 2027. These shifts make it practical to verify offline-to-online outcomes.

Evelyn observed, "Every untracked promo, print piece, vanity metric, stale report, and stalled test silently taxes our budget and deprives us of valuable insights. These overlooked wastes cost us money and prevent us from learning what truly works."

A Budget's Journey into the Void

Over the summer, Northbridge's sampling tour looked successful on the surface. Crowds engaged, photos were shared online, and local retailers applauded the extra foot traffic. While field teams were still on the road visiting key markets, analysts began compiling retailer feedback and social media activity. Yet when the finance team compared sell-through data, the uplift was negligible. The team had no mechanism to connect sample recipients to actual purchases. Without that link, the company could not learn which audiences the tour truly influenced, nor refine future tours for higher ROI. The hidden cost was not just the event's price tag; it was the lost learning that could have sharpened future campaigns.

Evidence sidebar: why QR now

Marketers report sharp momentum behind QR-enabled experiences and the broader transition to data-rich 2D codes: 93% increased QR use in the past 12 months, and global industry bodies are targeting **wide adoption of GS1-standard QR at retail by 2027**. These shifts make it practical to verify offline-to-online outcomes. [1]

Quantifying the Blind Spot: A Simple Ledger

To move from anecdotes to action, calculate four buckets.

1. **Unverified Spend Rate (USR):** The percentage of promotional budget spent on activities with no measurable customer outcomes or purchase verification.
   a. **USR** = Unverified promotion spend ÷ Total promotion spend.
   b. **Blind-spot tax** = USR × Total promotion budget.

2. **Decision Lag Cost (DLC):** The financial impact of delayed reporting that causes misallocation of media spend while waiting for verified results.
   a. **DLC** = (Average days to a verified read) × (Daily media spend at risk) × (% of spend likely misallocated).
   b. Benchmark to 7 days or less for high-velocity channels.

3. **Learning Debt (LD):** The lost optimization value from running tests without verification systems that could have confirmed performance improvements.

---

1. Bitly: From Scans to Strategy: How Marketers Use QR Codes in 2025 https://bitly.com/pages/qr-code-survey

a. **LD** = (# of tests run without verification) × (Estimated lift you could have confirmed) × (Campaign scale).
b. This is the value of optimizations you never banked.

4. **Inventory Misallocation Cost (IMC):** The margin loss from poor regional stock distribution due to a lack of real-time demand signals.
a. **IMC** = (Overstock units – OOS units avoided with better signals) × Unit margin.
b. Utilize store-level signals to optimize stock allocation by region.

Applied to Northbridge's situation:

- **USR:** 42% × $800K = $336K blind-spot tax
- **DLC:** 45-day delays × $12K daily spend × 30% misallocation = $162K quarterly loss
- **LD:** 8 unverified tests × missed 15% lift × $200K scale = $240K lost optimization
- **IMC:** 500 overstock units × $8 margin = $4K monthly inventory waste

Total Hidden Cost: $742K annually ($185K quarterly)

Calculating these hidden costs is only the first step. Reducing them requires coordinated action across departments, with each team taking ownership of specific metrics and inputs. The Hidden-Cost Ledger works best when responsibility is distributed based on natural organizational strengths and data access.

Cross-Functional Implementation:

- **Finance:** Calculate and track the four hidden cost metrics monthly
- **Marketing:** Implement verification mechanisms for all campaigns over

$10K

- **Sales:** Provide store-level performance data for inventory optimization
- **Operations:** Use verified demand signals for regional stock allocation

These four line items form your **Hidden-Cost Ledger**. Finance can own the math. Marketing and Sales own the inputs. Operations validates the inventory impact. Together, they convert "unknown" into a dollar figure that leadership can act on.

Breaking the Cycle with First-Party Data

Evelyn's solution was elegantly simple: attach verification to every promotional touchpoint. QR codes on sampling cups. Receipt uploads for rebates. Two-question surveys for insights. Within days, not months, Northbridge would know which tactics drove purchases and which audiences were most effective at conversion.

What changes operationally:

- **Verification at the edge.** Every marketing channel has at least one low-friction capture point.
- **Consent and quality.** All records are permission-based and validated.
- **Speed to signal.** Reporting could close in weeks, not months.
- **Attribution clarity.** Source- and store-level results decide the budget.

This same approach is possible for any brand willing to connect offline experiences to verified purchase outcomes.

Privacy Compliance Note: All data collection requires clear consent language. Sample text: "By scanning this code, you agree to receive promotional emails and allow us to track your purchase for product improvement. Unsubscribe anytime." See Chapter 11 for complete compliance protocols.

Implementation Sequence:

Moving from waste identification to verification systems requires a systematic rollout approach. Rather than attempting to fix all unverified spend simultaneously, successful organizations follow a proven sequence that builds capability while demonstrating quick wins to secure ongoing support.

1. Audit current promotions using the Hidden-Cost Ledger formulas
2. Identify the three highest-cost unverified activities
3. Design verification mechanisms for each (QR codes, receipt uploads, surveys)
4. Create consent language and privacy protocols
5. Launch pilot with one campaign
6. Measure results weekly and adjust
7. Scale successful approaches to remaining activities

Sidebar: Real-world example

Platforms such as Promolytics allow brands to connect offline promotional touchpoints with verified purchase data in real-time. For example, a QR code at an in-store tasting can direct consumers to a short survey and offer an instant rebate. As shoppers scan, answer two quick questions, and upload their receipt, Promolytics creates a unique identifier for each new customer in the database and links each response to actual purchase behavior. This allows for the documentation and updating of each touchpoint the consumer goes through during the marketing campaign, eliminating guesswork and delays. *This continuous profile building is what turns raw interactions into actionable intelligence across the organization.* Later, as customers encounter new campaigns from the brand, every additional touchpoint is linked back to their unique ID. That living record becomes the foundation for better decisions across Marketing, Sales, and Finance.

Marketing teams can instantly see which store events are the most effective. Sales can show retailers verifiable lift to negotiate prime placement, and Finance can reassign underperforming tactics before the budget is exhausted. By capturing and activating first-party data from the start, brands turn one-time promotions into ongoing customer relationships.

Early Results

- **Redemption rate:** Within the first two days of launch, 18% of tasters uploaded a receipt, indicating an immediate linkage between product trial and purchase. The analytics team segmented responses by store location and event date, pinpointing which tactics drove the highest engagement.

- **Incremental lift:** Stores participating in the QR-enabled sampling tour experienced a double-digit percentage increase in sales compared to nearby stores that did not host the event, directly attributing the uplift to the pilot.

- **Data collected:** 1,200 new consented customer contacts were gathered, each tagged with the taster's flavor preference and the timing of their purchase, enriching audience profiles for future targeted campaigns.

Lessons Learned

1. **Every dollar should teach you something**: Promotional spend is tuition; demand an education in return.

2. **Verification beats estimation**: Receipt data settles a debate quickly.

3. **Speed matters**: Rapid feedback loops allow mid-campaign pivots, protecting the remaining budget.

Executive Reflection – Budget Clarity:

Audit your last quarter's promotional spend using the Hidden-Cost Ledger framework. Which line items in your budget lack direct verification mechanisms? Calculate your organization's Unverified Spend Rate and estimate the annual cost of measurement gaps. What would reducing unverified spend by 25% mean for your team's learning velocity and budget efficiency? Identify three specific campaigns or activities where you could implement verification systems within the next 90 days.

Chapter 2 Checklist: Reducing Unverified Spend

- Calculate your organization's Hidden-Cost Ledger baseline using the four formulas.

- Assign cross-functional owners for each metric (Finance, Marketing, Sales, Operations).

- Inventory all active promotions and flag those without outcome tracking.

- Quantify the cost of unverified spend. See Appendix B for worksheets tailored to marketing managers and brand strategists.

- Implement at least one low-friction data capture mechanism (e.g., QR code, URL, or digital receipt) and confirm compliance with the privacy protocols outlined in Chapter 11.
- Set a concrete reduction goal (for example, cut unverified spend by 25% within two quarters) and assign an owner.
- Establish a cross-functional review cadence to monitor verified vs. unverified spend, reallocate budget based on proof, and plan the next test.

Verification solves the waste problem, but the real value emerges when you transform those verified signals into intelligent decisions. The next chapter demonstrates how to construct analytical frameworks that leverage customer data to gain a competitive advantage.

Chapter Three

The Signal Trap:

When Good Metrics Lead to Bad Decisions

Data-driven organizations face a hidden threat: metrics that look healthy while masking declining performance. This chapter reveals seven common ways that clean, accurate data can mislead decision-makers into making confident but incorrect choices.

We'll examine how correlation can be confused with causation, why small samples can lead to false confidence, and how digital-only dashboards can overlook critical offline signals. You'll learn to implement 'signal integrity' protocols that prevent metric misreads and build cross-channel measurement systems that reveal true business impact.

By the end of this chapter, you'll have practical tools for distinguishing meaningful signals from vanity metrics and frameworks for ensuring every dashboard drives profitable decisions.

The Confusing Dashboard

Northbridge's weekly dashboard painted a picture of marketing success, with a 47% increase in website sessions, email open rates reaching 28%, and social engagement up 65%. Yet when Evelyn cross-referenced these metrics with actual sales data, revenue had declined 8% month-over-month. The disconnect was jarring: how could every engagement metric surge while business results deteriorated?

To address this confusion, Evelyn gathered the team and walked through seven common deception patterns that had crept into their reviews. Misinterpretation does not always come from a lack of data. Often, it stems from misreading the wrong signals or misinterpreting the right ones without context.

Seven Ways Data Misleads – and How to Fix It

1. **Correlation conflation**
    - ***Trap:*** Two metrics move together and are assumed to be cause and effect.
    - ***Northbridge example:*** Coupon redemptions rose as repeat purchases rose. Seasonal demand was the true driver of both.
    - ***Fix:*** Use holdouts or geo-tests to estimate incrementality, which is the additional business outcome directly caused by your marketing activity, rather than what would have happened naturally. Prefer controlled experiments over observational reads, and align results to an Overall Evaluation Criterion (OEC), a single metric that best predicts long-term business value rather than short-term engagement. [1]
2. **Sample-size distortion**
    - ***Trap:*** Small pilots show dramatic swings that vanish at scale.
    - ***Northbridge example***: A banner variant showed a 40% lift over the weekend, then reverted to parity when rolled out.
    - ***Fix***: Pre-set minimum sample sizes and power thresholds before

1. EXP Platform: Pitfalls of Long-Term Online Controlled Experiments

scaling. Require region-level replication.

3. **Survivorship bias**
    - ***Trap*:** Only active customers are heard. Silent churners are invisible, so friction hides.
    - ***Northbridge example*:** Surveys showed high satisfaction, while repeat orders fell.
    - ***Fix*:** Proactively engage customers who have not had any activity in 90 days. Add exit surveys and follow-ups.
4. **Averages obscurity**
    - *Trap*: Means mask extremes and segment differences.
    - ***Northbridge example***: The average order value remained stable, while high-value customers spent less, and coupon hunters increased totals.
    - ***Fix*:** Break out metrics by lifecycle stage, cohort, and region. Track distribution, not just central tendency.
5. **Recency overweighting**
    - ***Trap*:** Short-term spikes are mistaken for durable trends.
    - ***Northbridge example*:** A competitor's weekend promo drove a dip in traffic and triggered overreaction.
    - ***Fix*:** Use rolling windows and control charts. Require a sustained signal before strategic changes.
6. **Algorithmic echo chamber**

- *Trap*: Recommenders and targeting continue to show more of what already performs, suppressing exploration..
- *Northbridge example*: Email continued promoting one top line, starving emerging categories.
- *Fix*: Allocate a fixed exploration budget and force diversity constraints in recommendations.

7. **Omnichannel myopia**
   - *Trap*: Dashboards overemphasize clicks while ignoring tastings, shelf talkers, displays, and store execution.
   - *Reasoning*: Most retail sales still occur offline, so digital-only views miss the outcome. In Q2 2025, U.S. e-commerce accounted for 15 .5% of retail, indicating that the majority of purchases still occur in physical channels that require offline verification.[2]
   - *Fix*: Integrate store-level sell-through, event feedback, and receipt data with digital metrics to enhance overall performance.

Misleading vs Meaningful Metrics: Separating Vanity from Value

To focus on true performance, Northbridge now compares traditional "look-good" metrics with outcome-focused alternatives that directly predict revenue or retention. Use this guide to highlight which metrics matter, and identify where to shift your team's analytics approach for more meaningful results.

---

2. United States Census, Quarterly Retail E-Commerce Sales Report, Aug 19, 2025

- **Website sessions** → prioritize **qualified sessions**, depth of visit, and verified conversions.

- **Email opens** → prioritize **click-to-delivered**, reply rate, and downstream purchase.

- **Social engagement** → prioritize **incremental sales** in matched regions and **retention** in exposed cohorts.

- **Click-through Rate (CTR)** → prioritize **lift vs. holdout** and **cost per incremental outcome.** Instead of celebrating raw click percentages, measure the additional clicks generated compared to a control group that didn't receive the campaign (a holdout group), and then calculate the true cost of each incremental action that wouldn't have occurred naturally.

- **QR scans** → prioritize **receipt-verified purchases** and **store-level sell-through** in the same time window.

- **Average order value** → prioritize **AOV by segment** and **customer lifetime value** estimates tied to verified identity.

- **Reach** → prioritize **reach of verified prospects** and **frequency capping** linked to incremental outcomes.

Restoring Context: Guardrails That Prevent Misreads

Evelyn instituted a simple protocol: every metric review:

- **Alternative explanations**. For any improvement, identify at least one benign interpretation and one risk interpretation.

- **Sample-size thresholds**. No scaling decisions will be made until the minimum sample and power requirements are met.

- **Silent churn checks**. Monthly outreach to inactive customers to surface friction missing from active feedback.

- **OEC first**. Tie every readout to the agreed Overall Evaluation Criterion (Appendix A) that predicts long-term value, not just short-term clicks or opens.

To account for silent churners, customer service began sampling past customers who had not ordered in ninety days and sent them a brief exit survey. Insights from this outreach quickly revealed packaging concerns that were absent from active customer feedback.

Cross-Functional Implementation

Preventing data deception requires a systematic approach to accountability across departments. Each team must own specific aspects of signal integrity based on their access to data sources and decision-making authority. Without clear ownership, even the best guardrails become suggestions rather than standards.

- **Marketing:** Implement bot filtering, set sample-size thresholds, and integrate offline event data

- **Sales:** Provide real-time store inventory and sell-through data

- **Finance:** Enforce minimum sample requirements before budget reallocation decisions

- **Operations:** Alert teams when promotional demand exceeds shelf availability

- **Analytics:** Run weekly data quality audits and survivorship bias checks

Northbridge Illustration: The "Summer Sip" Misread

The Summer Sip campaign illustrates how multiple deception patterns can compound. What appeared to be a 65% engagement spike was actually bot traffic (correlation conflation: assuming two metrics that move together have a cause-and-effect relationship), early results from one enthusiastic ambassador (sample-size distortion: drawing broad conclusions from unrepresentatively small test groups), and digital metrics that ignored empty store shelves (omnichannel myopia: overweighting digital signals while missing critical offline execution factors).

- **Week 1 dashboard:** Social mentions up 65%, QR scans soaring, site sessions at a three-month high.
- **Finance check:** Sell-through unchanged; retail re-orders dipped in two core regions.

What went wrong?

1. **Correlation conflation** – Scans and mentions rose together, but the cause was not shoppers.
2. **Bot inflation**. A scripting bot scraped the coupon page, and employees repeatedly tested links, inflating scans.
3. **Sample-size distortion** – Early uplift reports came from a single urban store, where an enthusiastic brand ambassador hand-sold bottles. The spike looked impressive but collapsed when rolled out nationally; the original sample was too narrow to represent the market.
4. **Omnichannel myopia** – Dashboards tracked digital interest, but the team failed to link intent signals with real-world execution. While dash-

boards tracked likes, clicks, and QR scans, sales teams weren't alerted to adjust orders or ensure displays were stocked. The result: consumers who were primed to buy found empty shelves, turning digital enthusiasm into missed revenue.

5. **Averages obscurity** – A blended engagement rate hid the fact that loyal buyers ignored the campaign while one-time coupon hunters spiked totals.

Course correction:

Evelyn paused new media spend and ran a 48-hour cross-functional audit. Analytics validated coupon redemptions against POS data, while the creative team reworked ad copy and visuals for regions showing genuine offline traction. The team:

- Matched coupon redemptions to POS data to verify actual incremental sales.
- Set bot-filtering rules on QR scan analytics.
- Added tasting-event feedback and regional sell-through to the core dashboard.
- Implemented a weekly demand-signal sync across Sales, Marketing, and Operations to anticipate promo-driven spikes and maintain on-shelf product availability.
- Re-forecasted ROI using only verified, cross-channel data.

Outcome

Within a month, by combining insights from both digital and in-store signals, Northbridge reallocated resources to where indicators genuinely aligned. Sales improved, demonstrating that success comes from connecting clean, contextual data with business outcomes rather than relying on superficial numbers.

Key takeaway: Linking metrics to outcomes ensures informed, effective actions.

Sidebar: Real-world example

Promolytics helps brands avoid the "false positive" trap that plagued Northbridge's Summer Sip campaign. By integrating QR surveys, receipt uploads, and offline event feedback into a single dashboard, we blend digital and in-store engagement metrics.

With Promolytics, teams verify engagement spikes against real sales, ensuring no metric is interpreted in isolation. The key takeaway: validated, unified data empowers smarter, mid-campaign decisions that protect ROI and directly improve results.

Practical Tooling: Your "Signal Integrity" Kit

Use this kit to refine your metrics, eliminating noise and lag, so budget decisions reflect verified outcomes. Assign an owner for each control, instrument adoption, and make integrity checks part of the weekly cadence.

- **Bot filters**: Rate limits, IP throttling, and CAPTCHA on promo pages. Track the share of traffic filtered.

- **Experiment design**: Utilize holdouts, geographically matched markets, or synthetic controls. Pre-register success metrics and minimum sample.

Align reads to an Overall Evaluation Criterion (OEC), a single metric that best predicts long-term business value (see Appendix A).

- **Cross-channel stitching**: Link QR scans, surveys, and receipts to store-level sell-through in the same time window.
- **Privacy Compliance:** Cross-channel data linking requires clear consent. Sample language: 'By participating, you consent to linking your responses with purchase data for product improvement.' See Chapter 11 for complete protocols.
- **Latency SLAs**: Set maximum time from event to verified read. Long reporting lags create bad decisions.
- **Exploration budget**: Reserve a fixed percent for testing new audiences, messages, and formats to avoid algorithmic lock-in.

Step-by-Step Signal Integrity Implementation

Recovering from data deception requires systematic intervention rather than ad-hoc fixes. Organizations that successfully eliminate metric misreads follow a structured four-phase approach that builds capability while demonstrating quick wins. Each phase has specific deliverables and timelines designed to prevent the overwhelming "boil the ocean" approach that often stalls implementation.

Phase 1: Immediate Assessment (Week 1)

- **Audit current metrics** against the seven deception patterns outlined above
- **Freeze questionable spending** on campaigns that show suspicious engagement spikes

- **Convene a cross-functional team** to diagnose metric discrepancies within 48 hours

Phase 2: Data Validation (Week 2)

- **Implement bot filtering** using IP throttling and CAPTCHA on landing pages
- **Validate redemptions** by mapping each coupon use to point-of-sale data
- **Cross-reference engagement** with store-level sell-through in the same time windows

Phase 3: Infrastructure Hardening (Weeks 3-4)

- **Set sample-size guardrails** (minimum 1,000 verified engagements per region)
- **Integrate offline data** (tastings, retail feedback) into the core dashboard
- **Break out segment-specific metrics** by lifecycle stage to surface hidden patterns

Phase 4: Operational Changes (Ongoing)

- **Establish weekly data quality reviews** with the cross-functional team
- **Reallocate budgets** based on verified ROI rather than vanity metrics
- **Document assumptions** in every dashboard to prevent future misreads

This approach:

- **Creates logical progression** from assessment to implementation
- **Sets realistic timelines** that executives can commit to
- **Groups related activities** into manageable phases
- **Provides measurable milestones** (48-hour audits, 1,000-engagement minimums)
- **Eliminates the narrative elements** that don't belong in implementation guidance

Executive Reflection:

Audit your organization's core performance dashboards against the seven deception patterns outlined in this chapter. Which metrics are you celebrating that lack verified business outcomes? Where might correlation be masking true causation in your decision-making? Identify three specific metrics your team tracks that need context or verification to prevent misreads within the next 30 days.

Chapter 3 Checklist: Preventing Data Deception

- Link every headline metric to verified business outcomes (sales, retention, margin).
- Identify alternative explanations for any two metrics that move together.
- Set minimum sample-size thresholds before scaling decisions (1,000+ events recommended).
- Segment metrics by customer lifecycle, source, and region instead of

using averages.

- Filter bots and internal traffic from engagement metrics using IP and CAPTCHA controls.
- Integrate offline data (tastings, store sales, events) with digital analytics.
- Survey inactive customers quarterly to identify hidden friction points.
- Establish weekly cross-team data quality reviews to catch anomalies early.

Signal integrity protocols address internal measurement issues, but most organizations still heavily rely on external data sources that introduce their own blind spots. Industry reports, syndicated research, and third-party audience profiles create a different kind of deception, one that feels authoritative while masking critical gaps in customer understanding. The next chapter reveals how third-party data dependence can create false confidence and demonstrates how to develop superior competitive intelligence through the use of first-party alternatives.

Chapter Four

The Dependency Dilemma:

Hidden Costs of External Data Reliance

Third-party data—customer information purchased from external vendors rather than collected directly from your own audience interactions—creates an illusion of scale while introducing hidden risks that compound over time.

We'll examine the structural limitations of rented audiences, introduce Risk-Adjusted Incremental ROAS (RA-iROAS)—a metric that factors data quality, compliance risk, and supply stability into traditional return calculations—for evaluating external data sources, and provide a managed drawdown strategy for shifting budget from third-party segments to owned cohorts without losing acquisition volume."

By the end of this chapter, you'll have frameworks for auditing third-party dependencies, tools for building superior first-party alternatives, and a practical transition plan that protects performance while building sustainable competitive advantage."

Northbridge: Rent or Own

Later that month, Evelyn and Malik attended a planning session with their media agency, while the analytics team continued to monitor results from the recent first-party pilots, experiments using data collected directly from their own customers. The agency recommended increasing spend on third-party audience segments, which are customer groups based on data purchased from outside vendors, to counter slowing sales. The proposal looked polished, but Evelyn hesitated. Over the past year, performance from these external audiences had grown increasingly inconsistent. As their first-party data pilots began to show promise, she questioned whether increasing dependence on external targeting sources was the right move.

Why Third-Party Data Creates Hidden Vulnerabilities

Short take. Third-party audiences feel like a shortcut: instant scale, labeled consumers, quick reach. In practice, they introduce inconsistency, opacity, compliance risk, and competitive parity. The fix is a managed drawdown plan that replaces rented reach with owned, verified cohorts.

Five Critical Weaknesses of External Audiences

1. **Relevance**: External segments are broad. They rarely reflect the specific motivations of Northbridge's flavored beverage and snack buyers. Generic labels obscure important nuances, such as bundle behavior or regional preferences.

2. **Freshness**: Update cycles lag behind real-time shifts. By the time segments refresh, Northbridge may already notice changes through receipts or surveys.

3. **Control**: Methodology changes at the data provider break continuity. Historical comparisons become unreliable. Northbridge experienced this when a lookalike product definition was altered, invalidating prior benchmarks.

4. **Compliance risk**: Increased US privacy scrutiny demands clear consent. Opaque sourcing raises legal and reputational risks.

5. **Competitive parity**: If competitors buy the same audiences, differentiation relies only on creative or pricing. Lasting advantage fades.

The Mounting Risks of Third-Party Reliance

Third-party audiences appear stable until policy shifts, vendor changes, and signal pollution reveal how little control you have. This section summarizes the operational, legal, and performance risks that make third-party dependency fragile and shows how those risks distort planning, ROI, and accountability.

- **Policy volatility.** Chrome's third-party cookie timeline has been in flux under UK regulator oversight, underlining that dependency on cross-site tracking is a strategic risk, regardless of timing.[1]

- **State privacy laws are widening.** New and expanding U.S. state laws (for example, Texas TDPSA effective July 1, 2024, with ongoing obligations) increase the bar for consent and data handling. Opaque sourcing raises exposure.[2 3 4]

- **Signal pollution.** Automated traffic now accounts for a significant share of the web; bad bots alone comprise ~37% of traffic, inflating scans,

1. CookieYes, Third-Party Cookies Going Away? Here's What's Actually Happening, May 28, 2025

2. Attorney General of Texas, Texas Data Privacy and Security Act, effective July 1, 2024

3. iapp25 Resource Center, US State Privacy Legislation Tracker

4. Bloomberg Law, Which States Have Consumer Data Privacy Laws? April 7, 2025

5. Thales Cybersecurity, 2025 Bad Bot Report

clicks, and sessions unless filtered.[5]

- **Segment inaccuracy.** Independent research reveals that brokered audience segments are often inaccurate or outdated, undermining claims of target quality and lift.[6]

Dependency Assessment: Three Critical Questions

A quick self-check reveals hidden exposure. Use these questions to confirm whether you can quickly replace a rented segment, prove clean provenance, and trust your reads. Any "no" flags a priority fix for the next buying cycle.

Answer yes/no:

1. If a key third-party segment were to vanish tomorrow, could you replace it with an owned cohort within 30 days?
2. Do you know the provenance and last refresh date of consent for every rented segment?
3. Do your dashboards filter bot traffic and separate MPP-affected email opens from decision metrics?

If any answer is "no," prioritize dependency reduction before the next buying cycle.

Risk-Adjusted ROAS: True Cost of External Data

---

5. Thales Cybersecurity, 2025 Bad Bot Report

6. informs PubsOnLine, Frontiers: How Effective is Third-Party Consumer Profiling? Evidence from Field Studies, Nico Neurmann, Catherine E. Tucker, Timothy Whitfield, Oct 2, 2019

Move beyond raw ROAS. Use **Risk-Adjusted Incremental ROAS (RA-iROAS)** that prices in quality, compliance, and supply volatility.

- **Incremental ROAS (iROAS)** = Incremental Revenue ÷ Media Spend
- **Quality discount** = 1 – (Invalid Traffic %) × (Verification gap factor)
- **Provenance penalty** = 1 – Compliance confidence score (0–1)
- **Supply volatility penalty** = 1 – Probability the segment remains stable over your planning window

RA-iROAS = iROAS × Quality discount × Provenance penalty × Supply volatility penalty

Applied to Northbridge's 'family value' segment:

- iROAS: 3.2x
- Quality discount: 0.82 (18% invalid traffic)
- Provenance penalty: 0.75 (medium compliance confidence)
- Supply volatility penalty: 0.90 (10% chance of segment changes)
- RA-iROAS: 3.2 × 0.82 × 0.75 × 0.90 = 1.77x

Set an **RA-iROAS floor** for any third-party segment to stay live. If a segment fails, it is assigned to either "sunset or fix" with a dated remediation plan.

Contract Controls for Vendor Accountability

Strong contracts reduce surprise. Use these guardrails in briefs and MSAs to require provenance warranties, change-notice SLAs, audit rights, and clear remedies so vendor supply remains compliant, stable, and aligned with your performance standards.

- **Provenance warranties**: lawful basis, consent records, and right to audit on request.
- **Change-notice SLA**: 90-day notice for methodology changes or segment retirement.
- **Data minimization**: no sensitive-location or sensitive-attribute targeting; provider certifies exclusion.
- **Indemnity & Clawbacks:** Clawback fees for non-conforming supply; indemnify against privacy claims related to provider sourcing.
- **Performance warranty**: if incrementality < threshold for 2 consecutive windows, automatic pause or price adjustment.

Cross-Functional Implementation

Reducing third-party dependence requires coordinated action across departments, with each team taking ownership of specific risk assessment and transition activities based on their expertise and decision-making authority. Without clear accountability, even comprehensive frameworks become suggestions rather than systematic change.

- **Marketing:** Execute side-by-side tests and manage drawdown timeline
- **Legal:** Audit vendor contracts and compliance documentation
- **Finance:** Track RA-iROAS metrics and budget reallocation
- **Analytics:** Implement bot filtering and verification systems

Mapping Your External Data Dependencies

Before reallocating the budget, make the dependency visible. This inventory captures the provenance, consent basis, refresh cadence, stability, and incremental performance of each segment, enabling Finance, Legal, and Marketing to assess risk, compare options, and determine what to retain, modify, or sunset.

Create a one-pager per segment:

- **Source & method.** How collected, consent basis, refresh cadence, and exclusivity. Require written attestations.
- **Match and reach.** On-target rate, overlap with your owned IDs, and uniqueness compared to competitors.
- **Stability.** 12-month deprecation history, rename frequency, API/ID dependencies.
- **Performance.** Incremental lift vs. matched owned cohort; RA-iROAS vs. floor.
- **Risks.** Jurisdictions affected by state privacy laws; sensitive data flags

First-Party Advantage: Four Pillars of Owned Audiences

Lasting advantage comes from audiences you capture with consent and can enrich over time. This section defines what "good" looks like for owned cohorts, including identity, freshness, unique signals, and activation patterns that compound with every touchpoint.

- **Identity with consent.** QR code + short survey + receipt validation creates durable, permission-based profiles you control.

- **Freshness by design.** Real-time enrichment from events, POS, and service logs beats quarterly broker refreshes.
- **Unique signals.** Bundle affinity, flavor preference, store-of-purchase, and timing windows are hard to buy and easy to use.
- **Compounding advantage.** Owned cohorts become more predictive with each touchpoint, unlike rented segments that tend to decay.

Privacy Compliance Note: All first-party capture requires clear consent language. Sample text: 'By participating, you agree to receive promotional communications and allow us to track your purchase for product improvement. Unsubscribe anytime.' Ensure compliance with applicable state privacy laws, including CCPA, CPRA, and emerging regulations. See Chapter 11 for complete protocols.

Northbridge's Reality Check

During the 2024 holiday season, Northbridge allocated $240K to a third-party "family value" segment targeting households with children aged 6-14. The external audience promised 890K reach and delivered impressive initial metrics: 2.3% Click-Through Rate, 47K landing page visits, and 12K coupon downloads.

Yet when finance cross-referenced the campaign with actual sales data, the results were sobering. Incremental revenue totaled just $180K, a 0.75x ROAS that fell well below Northbridge's 2.5x minimum threshold. Store managers reported that many coupon users were unfamiliar with the brand, indicating that the segment primarily attracted deal hunters rather than genuine prospects.

Meanwhile, a parallel first-party pilot told a different story. Using QR codes at in-store tastings, Northbridge gathered 3,200 consented emails from families actively sampling their products. These contacts received targeted bundle offers based on their stated flavor preferences. Despite reaching only 22% of the third-party segment's volume, the first-party cohort generated $195K in incremental revenue, a 3.1x ROAS.

The contrast was stark:

- **Third-party segment:** $240K spend, 0.75x ROAS, 8% repeat purchase rate
- **First-party cohort:** $63K spend, 3.1x ROAS, 34% repeat purchase rate

The third-party "family value" label had obscured a critical insight: parents willing to try new products with their children represented far higher intent than households that simply matched demographic criteria. The owned cohort didn't just outperform on immediate returns; it built relationships that compounded over time.

This revelation prompted Northbridge to redirect $85K from the underperforming external segment to expand its tasting program, ultimately generating an additional $140K in verified incremental sales while building 2,800 new first-party customer relationships.

Key insight: Behavioral signals from actual product interaction out predict demographic assumptions every time.

Sidebar: Real-world example

Promolytics lets any brand capture, verify, and activate its own audiences, as Northbridge did. It helps brands reduce their dependence on third-party data by building their own audiences. Promotions, receipt validation, and surveys capture usable, consented data instantly.

Instead of relying solely on a purchased "family value" segment, a brand can run a display-based campaign to collect first-party contact information (email, social media, SMS), proof of purchase, and stated interest in bundles.

Owned cohorts consistently outperform rented audiences, delivering higher repeat purchases and lower churn rates, while remaining privacy-compliant.

Head-to-Head Testing: Owned vs. External Segments

Debates end when comparable cohorts are tested under the same rules. This design outlines how to pit a rented segment against an owned cohort using a pre-registered Overall Evaluation Criterion (OEC), as outlined in Appendix A, matched markets, verified outcomes, and minimum sample thresholds, ensuring the winner is unambiguous.

- **Define the OEC upfront** (for example, receipt-verified incremental sales per exposed shopper within 28 days, net of promo cost).
- **Set markets**: pick matched regions; cap spend equally.
- **Verification**: link exposure to **POS or receipt uploads**; filter invalid traffic.
- **Minimums**: pre-register sample sizes, analysis window, and guardrails.
- **Decision rule**: shift budget only if the owned cohort outperforms the third-party segment on the OEC and the guardrails are met.

Managed Transition: From Rented to Owned Audiences

Shifting from third-party dependence to first-party advantage requires a systematic approach rather than sudden changes that risk compromising acquisition volume. Organizations that successfully complete this transition follow a structured approach that builds owned audience capability while maintaining performance throughout the drawdown period.

Strategic Framework

The transition balances three priorities: protecting current acquisition volume, building superior owned alternatives, and reducing dependency risk. Rather than choosing between control and growth, successful brands employ a phased approach that demonstrates owned cohort superiority before reallocating a significant budget.

Core Principles:

- **Evidence-driven reallocation:** Budget shifts only after owned cohorts demonstrate superior performance
- **Risk management:** Gradual drawdown prevents sudden volume drops
- **Capability building:** Each quarter focuses on strengthening first-party capture and activation

Four-Quarter Transition Plan

Quarter 0: Foundation Phase

- Inventory all segments; baseline RA-iROAS; stand up capture points to grow owned IDs

Quarter 1: Initial Testing (15-25% budget shift)

- Shift 15-25% of spend to owned cohorts
- Run matched-market tests
- Set provenance SLAs with vendors

Quarter 2: Scaling Winners (35-50% budget shift)

- Lift the shift to 35-50% where owned cohorts win
- Sunset third-party segments that fail RA-iROAS minimum thresholds
- Increase first-party capture investment in channels showing strong consent rates
- Document learnings for creative and offer optimization

Quarter 3: Optimization Phase

- Consolidate budget around top-performing owned cohorts
- Keep a 10-15% exploration budget for contextual and partner data that passes provenance and performance gates
- Implement automated decision rules for future third-party segment evaluation
- Establish ongoing quarterly dependency reviews

Implementation Guardrails

Volume Protection:

- Set minimum acquisition thresholds before any budget reallocation
- Use matched-market testing to isolate performance differences
- Maintain emergency reversion protocols if owned cohorts underperform

Quality Assurance:

- Verify that all first-party capture includes proper consent language
- Implement bot filtering across all owned audience development
- Cross-reference engagement with verified purchase behavior

Performance Measurement:

- Use consistent OEC across all cohort comparisons
- Pre-register test parameters and success criteria
- Document methodology changes to maintain historical comparability

Northbridge Results

Following this framework, Northbridge reduced its share of third-party spend while maintaining acquisition volume. Owned cohorts exhibited higher repeat purchase rates and lower churn signals. Leadership gained confidence that dependence could continue decreasing without jeopardizing short-term goals.

The systematic approach has proven that dependency reduction enhances rather than compromises marketing effectiveness when executed with proper measurement and gradual implementation.

Executive Reflection – Dependency Risk:

Audit your organization's third-party data dependencies using the inventory framework outlined in this chapter. Which external audience segment would most significantly disrupt your marketing performance if it were removed tomorrow? Calculate the RA-iROAS for your top three third-party segments and identify which owned cohorts could serve as alternatives. What percentage of your current audience targeting budget could you realistically shift to first-party capture within two quarters?

Chapter 4 Checklist: Reducing Third-Party Reliance

- Inventory all third-party segments with provenance documentation.
- Calculate RA-iROAS for top external audience sources.
- Set target percentage for budget shift to first-party capture (15-25% Quarter 1).
- Launch one side-by-side test comparing owned vs. rented cohorts.
- Implement vendor contract guardrails for compliance and performance.
- Establish a quarterly dependency review with Legal and Privacy teams.

Reducing spend on third-party data gave Northbridge breathing room, but it also exposed a fundamental challenge: without a unified framework for first-party data collection and activation, progress risked fragmenting across disconnected pilots. The team had proven that owned cohorts could deliver superior outcomes, yet their customer signals remained scattered across receipts, surveys, service logs, and promotional touchpoints, valuable but uncoordinated.

Evelyn recognized that replacing rented audiences with a sustainable competitive advantage required more than successful tests. Northbridge required systematic foundations, including clear definitions of what constitutes quality first-party data, unified collection standards across all customer touchpoints, and activation protocols that transformed scattered signals into coordinated customer intelligence.

The next challenge was building those foundations without disrupting the momentum they had already created. Success would require establishing first-party data standards that every department could implement consistently while scaling the capture mechanisms that had proven most effective in their transition away from third-party dependence.

Chapter Five

First-Party Fundamentals:

When Consent-Based Collection Powers Strategic Advantage

First-party data is the only signal set you fully control; however, defining it properly determines whether you build a competitive advantage or create expensive confusion. This chapter establishes what qualifies as legitimate first-party data, how to capture it with proper consent, and how to make it decision-grade (data that meets quality standards for driving business decisions), through accuracy, relevance, and freshness standards.

We'll examine the critical distinctions between first, second, and third-party data sources, introduce quality dimensions that ensure reliability, and provide frameworks for cross-functional activation. You'll learn to transform scattered customer touchpoints into unified intelligence that drives Marketing, Sales, Product, and Finance decisions.

By the end of this chapter, you'll have clear definitions, quality standards, and governance protocols that turn every customer interaction into verified business intelligence.

Northbridge: From Definition to Deployment

Early in the next planning cycle, Evelyn meets with a cross-functional working group comprising Marketing Operations, Data Engineering, Sales, Customer Service, and Product Development, while the finance team finalizes budget reallocations based on recent campaign learnings. The goal is clear: move Northbridge from recognizing its blind spots to building a durable first-party foundation. Before tools or dashboards, the team needed a shared definition of what exactly constitutes first-party data and how it will be utilized.

What Is First-Party Data

First-party data is information customers share directly through their interactions with your brand, both explicit (survey responses, account details) and implicit (purchase history, browsing patterns), always within a clear consent relationship. A consent relationship means customers knowingly provide information in exchange for value, whether that's product improvements, personalized offers, or better service. Unlike external data sources that tell you about generalized market segments, first-party data reveals what your actual customers do, when they do it, and why. (For more detailed information on Data Party levels, see Appendix C · Data Source Comparison Guide)

Quality Dimensions

1. **Accuracy**: Verified at the point of collection through receipt uploads, confirmed purchases, or validated responses

2. **Relevance**: Directly connected to your products, services, and customer journey stages

3. **Freshness**: Updated in real-time or near real-time as customer behavior evolves

Integrity across these dimensions makes the dataset trustworthy enough to drive decisions.

The discussion moved from theory to practice with Northbridge's upcoming launch, an opportunity to see first-party data in action.

Implementing these quality standards requires careful attention to legal requirements and customer rights.

Privacy Compliance: All first-party collection requires transparent consent language. Example: 'By providing this information, you agree to receive product updates and allow us to improve your experience. Opt-out anytime.' Ensure compliance with CCPA, CPRA, and applicable state privacy laws. See Chapter 11 for detailed protocols.

Cross-Functional Implementation

Establishing first-party data standards requires coordinated execution across departments, with each team taking ownership of specific collection, validation, and activation responsibilities. Without clear accountability for data quality and usage protocols, even well-defined standards become inconsistent suggestions rather than systematic competitive advantages.

- **Marketing:** Design consent-compliant capture mechanisms across campaigns and touchpoints
- **Sales:** Provide customer interaction data and purchase verification from retail channels
- **Product:** Use verified customer feedback for development priorities and feature roadmaps
- **Data Engineering:** Ensure quality standards, integration protocols, and real-time data processing
- **Legal:** Maintain consent documentation, compliance oversight, and privacy protocol updates

Northbridge Illustration: First-Party Data in Action

During Northbridge's spring citrus blend launch, the team deployed a coordinated first-party capture strategy across three channels. QR-enabled shelf talkers in 47 retail locations directed shoppers to a 30-second survey offering an instant $1.50 rebate. Social media ads drove traffic to the same survey for online purchasers. In-store sampling events collected additional feedback through tablet surveys.

Week 1 Results:

- 2,847 QR scans generating 1,923 completed surveys (67% completion rate)
- 1,456 receipt uploads validated actual purchases (76% validated purchases)
- Including 812 sampling event responses with immediate purchase verification

Key Data Quality Indicators:

- 94% accuracy rate (verified against POS data)
- 100% consent compliance (clear opt-in language)
- Real-time integration (survey to CRM or database within 2 hours)

Actionable Insights Discovered: Flavor Preference Correlation - Customers selecting "citrus-forward" taste preferences showed a 3.2x higher repeat purchase rate within 30 days compared to those with "mild citrus" preferences.

Bundle Opportunity: 73% of citrus preference respondents also purchased complementary snack items, revealing a $4.80 average basket uplift when products were cross-merchandised.

Regional Performance Variation: Northeast markets showed a 40% higher conversion rate from survey to purchase compared to Southeast markets, indicating that different promotional strategies were needed.

Cross-Functional Activation:

- **Product Team:** Prioritized citrus-forward formulations for the next seasonal launch based on repeat purchase correlation
- **Sales Team:** Negotiated expanded shelf space in Northeast regions using verified demand data, securing 23% more facings
- **Marketing Team:** Shifted $35K from broad demographic targeting to retargeting the 847 high-intent citrus preference customers
- **Operations:** Adjusted inventory allocation, reducing Northeast stockouts by 18% while preventing Southeast overstock

Business Impact (90 days):

- 34% increase in repeat purchases among surveyed customers vs. the control group
- $127K incremental revenue attributed to data-driven decisions
- 2,156 new opted-in customer profiles added to CRM
- 89% customer retention rate among survey participants

Data Compound Effect: Six months later, this first-party cohort continued to outperform traditional acquisition campaigns by 2.4 times on customer lifetime value, demonstrating that behavioral signals from product interaction create sustainable competitive advantages over demographic assumptions.

Sidebar: Real-world example

Promolytics turns everyday interactions into structured, high-quality first-party data. Through links, first-party cookies, and QR-enabled touchpoints across tastings, print, media, and packaging (see Appendix E: QR Code Statistics & Placement Playbook), the platform captures consent, geolocation, timestamps, receipt-verified purchases, and stated preferences in real time. The result is a living consumer journey map that connects each touchpoint to its outcomes.

For example, imagine a beverage company launching a new regional flavor. The company places QR codes on in-store display signage, directing shoppers to a brief survey and an instant rebate offer. When shoppers scan the code, complete the survey, and upload their receipts, Promolytics collects detailed response data—such as the purchase date, time, channel, and store location—in real-time. Marketing teams can then use this data to retarget buyers who have shown high intent. Sales can provide retailers with proof of regional sales lift, and Product Development can refine future launches using verified consumer feedback, all without depending on third-party data.

By unifying capture, validation, and activation on one platform, Promolytics ensures that first-party insights aren't only collected but also immediately actionable.

Executive Reflection:

Audit your organization's current customer touchpoints and identify which interactions generate first-party data versus which rely on external sources. Where are the gaps between departments in how first-party data is defined, collected, and used? Which three customer interactions could be enhanced to capture higher-quality first-party signals within the next 60 days?

Chapter 5 Checklist: Building a First-Party Foundation

- Ensure leadership shares a unified definition of first-party data.
- Audit existing first-party sources (CRM, receipts, surveys, loyalty programs, and service logs) to identify available data types and potential gaps.
- Define quality dimensions and expected outcomes for accuracy, relevance, and freshness.
- Develop governance rules for the collection, storage, and tracking of consent.
- Integrate one pilot dataset (e.g., survey + POS) across teams to test cross-functional data use.
- Correlate insights against real outcomes (e.g., repeat purchase, basket value).
- Document learnings and standardize processes to ensure that improvements are consistently applied before scaling to new launches.

Defining first-party data standards provided Northbridge with clarity, but collecting data without integration creates new silos. While marketing celebrated digital engagement metrics and data engineering refined first-party capture systems, these insights remained disconnected, valuable individually but failing to create a comprehensive understanding of the customer.

The next challenge was building bridges between first-party signals and existing digital analytics to create unified customer intelligence that would amplify both data sources rather than forcing teams to choose between them.

Chapter Six

Amplifying Insight:

Merging First-Party Data with Digital Analytics

Digital analytics tell you who looked and where they came from. First-party data tells you who bought, why they bought, and where they made the purchase. When you merge these data sources, you transition from activity reports to decision-grade insights that link awareness to verified revenue, guiding budget allocation, creative optimization, and retail execution.

This chapter reveals how to integrate digital engagement metrics with first-party purchase verification to eliminate the gap between vanity metrics and business outcomes. We'll examine why parallel data streams create blind spots, introduce frameworks for connecting digital touchpoints to verified purchases, and provide step-by-step protocols for cross-validating campaign performance. Cross-validation confirms digital engagement metrics against verified purchase data to ensure accuracy and reliability.

By the end of this chapter, you'll have practical tools for merging digital analytics with first-party data, frameworks for identifying which channels actually drive revenue, and systematic approaches for reallocating budget based on integrated intelligence.

Northbridge: Seeing the Whole Picture

Evelyn sat at her desk, comparing two dashboards. On one screen, Google Analytics showed impressive website traffic spikes. On the other hand, the first-party data pilot revealed receipts uploaded from in-store promotions. The problem? The two stories didn't line up. While website traffic soared, only a fraction of visitors converted to purchases. It was only when she merged the datasets that the real truth came into focus: a regional TikTok influencer had driven a burst of online attention, but the true buyers were coming from a targeted rebate offer placed in grocery stores. Without merging these two perspectives, Northbridge would have celebrated digital "buzz" while overlooking the channel that actually delivered revenue.

Why Integration Creates Complete Intelligence

Activity is not an outcome. Digital dashboards report attention—impressions, clicks, visits, engagement rates. First-party data provides verified outcomes, including purchases, preferences, consented profiles, and actual revenue. Alone, each view is incomplete and can mislead decision-making.

The Blind Spot Problem:

- **Digital-only view:** Teams celebrate high engagement while missing conversion failures
- **First-party-only view:** Teams see purchases but miss the awareness drivers that enabled them
- **Integrated view:** Teams understand which digital activities actually translate into verified sales

Strategic Pairings That Reveal Truth:

- **Web + First-Party:** Match QR scans or rebate uploads with web sessions to identify which digital visitors actually purchase
- **Social + First-Party:** Align engagement spikes with verified redemption data to confirm which creative formats drive sales
- **CRM + First-Party:** Combine loyalty program engagement with purchase history to identify true high-value customers
- **Ad Platforms + First-Party:** Validate media spend by tying impressions to receipt-confirmed buyers, not estimated conversions

Integration reframes the fundamental question from "What happened online?" to "How did online and offline activity translate into verified sales and long-term customer value?" This shift from activity measurement to outcome verification enables confident budget allocation, creative optimization, and retail execution decisions based on proven performance rather than hopeful assumptions.

Northbridge Illustration: The Integration Breakthrough

During Northbridge's summer sparkling water launch, the marketing team deployed a $180K campaign across TikTok, Instagram, and in-store promotions. Traditional digital-only analysis would have declared victory based on impressive engagement metrics; however, an integrated analysis revealed a more complex and actionable truth.

Digital-Only View (Week 2):

- TikTok ads: 2.3M impressions, 4.8% engagement rate, 47K clicks
- Instagram campaigns: 1.8M reach, 6.2% engagement, 31K profile visits

- Estimated conversion rate: 2.1% (platform attribution)
- Apparent campaign ROAS: 3.2x

Integrated Analysis (Digital + First-Party Data): When Northbridge overlaid QR scan data and receipt uploads from in-store promotions, patterns emerged that digital analytics missed:

Cross-Validation Results:

- Only 23% of digital engagement correlated with verified purchases
- TikTok drove awareness, but Instagram users showed 3.7x higher purchase rates
- 67% of verified buyers had engaged with digital content within 48 hours before purchase
- True campaign ROAS: 4.8x (verified revenue attribution)

Regional Performance Disparities:

- **Northeast markets:** Digital engagement matched purchase verification (strong correlation)
- **Southeast markets:** High digital engagement but low conversion (awareness without action)
- **West Coast:** Moderate digital metrics but the highest per-customer purchase value

Strategic Pivots Based on Integration:

1. **Budget Reallocation:** Shifted $45K from TikTok to Instagram based on verified conversion data
2. **Regional Targeting:** Increased Northeast spend by 40% where digital correlated with purchases
3. **Creative Optimization:** Scaled Instagram creative formats that showed verified purchase lift
4. **Retail Execution:** Prioritized in-store display placement in regions with strong digital-to-purchase correlation

Measurable Business Impact (30 days):

- **Revenue increase:** $147K incremental sales attributed to integration insights
- **Efficiency gain:** 34% improvement in cost per verified acquisition
- **Customer intelligence:** 3,200 new first-party profiles with verified purchase history
- **Retail leverage:** Used verified lift data to secure 28% more shelf space across key markets

The Integration Advantage: Six months later, campaigns using integrated measurement consistently outperformed digital-only campaigns by 2.1x on verified ROAS. The team discovered that digital engagement without purchase verification led to overinvestment in awareness channels that failed to drive revenue, while integration revealed the precise combination of digital touchpoints and retail execution that converted browsers into buyers.

Key Insight: Digital metrics alone would have led to continued investment in TikTok, despite poor conversion performance. Integration revealed Instagram as the superior revenue driver, enabling budget optimization that generated an additional $147K in verified sales within 30 days.

Sidebar: Real-World Example

Promolytics helps brands connect the dots between **digital signals** and **first-party data**, enabling them to see when campaigns truly work in tandem. For example, a beverage company running TikTok and Instagram ads for a seasonal launch can use Social Media to track clicks and engagement, while Promolytics is simultaneously capturing survey responses and receipt uploads from shoppers who redeem in-store offers.

Once these two streams of data are merged, the brand moves from simply knowing which ads reached people to identifying which ads drove actual purchases, where these purchases occurred, and what consumers thought about the product. First, marketing pinpoints top-performing creatives, then sales demonstrates lift to retailers, and finally, product teams receive feedback to inform the next launch.

The result is a campaign that combines the **scale of digital analytics** with the **certainty of first-party data**, transforming surface-level engagement into verified, revenue-driving insights.

Framework for Merging First-Party and Digital Data

Transform parallel data streams into unified customer intelligence using this systematic five-phase approach. Each phase includes specific tools, timelines, and success criteria to ensure reliable integration and actionable outcomes.

Phase 1: Collect (Week 1)

- **Digital tracking:** Implement UTM parameters, pixel tracking, and platform APIs for all campaigns
- **First-party capture:** Deploy QR codes, URL links, first-party cookies, receipt uploads, and survey mechanisms at key touchpoints
- **Identity anchors:** Establish email, phone, or loyalty ID as the primary linking mechanism
- **Success metric:** 90%+ data capture rate across both digital and first-party touchpoints

Phase 2: Connect (Week 2)

- **Identity matching:** Use hashed email addresses or customer IDs to link digital sessions with purchase records
- **Time window rules:** Set windows for connecting digital engagement to purchases
- **Data validation:** Verify match rates exceed 60% for reliable cross-platform analysis
- **Technical setup:** Configure API connections between analytics platforms and CRM systems

Phase 3: Compare (Week 3)

- **Cross-validation testing:** Run holdout groups to verify digital attribution against actual purchase data

- **Noise elimination:** Filter bot traffic, internal sessions, and duplicate records from analysis
- **Statistical significance:** Requires a minimum sample size before drawing conclusions
- **Performance gaps:** Identify channels with high digital engagement but low verified conversion

Phase 4: Calibrate (Week 4)

- **Budget reallocation:** Shift 15-25% of spend toward channels with verified purchase correlation
- **Creative optimization:** Scale ad formats and messages that show cross-validated performance
- **Channel prioritization:** Rank touchpoints by verified ROI rather than estimated conversions
- **Regional adjustments:** Modify geographic targeting based on integrated performance data

Phase 5: Iterate (Ongoing)

- **Weekly reviews:** Analyze new integrated data for emerging patterns and optimization opportunities
- **Quarterly assessments:** Evaluate overall integration effectiveness and expand successful approaches
- **Feedback loops:** Feed verified conversion insights back into campaign targeting and creative development

- **Capability building:** Train teams on integrated analysis to reduce dependence on single-source metrics

Northbridge Application Example: During their summer launch, Northbridge followed this framework precisely. Phase 1 captured 2,847 QR scans and 94K digital interactions. Phase 2 connected 67% of digital users to verified purchases through device ID matching. Phase 3 revealed that Instagram users converted 3.7x better than TikTok users. Phase 4 reallocated $45K toward proven performers. Phase 5 established weekly integration reviews that consistently improved campaign performance.

This systematic approach ensures integration becomes a repeatable competitive advantage rather than a one-time analysis exercise.

Privacy Compliance: Cross-platform data integration requires consistent consent across all touchpoints. Ensure customers understand how their digital behavior and purchase data will be connected. Sample language: 'We combine your online activity with purchase history to improve your experience.' See Chapter 11 for complete protocols.

Cross-Functional Implementation

Integrating digital analytics with first-party data requires coordinated data flows and shared measurement standards across departments. Each team must contribute specific data sources and validation methods to create comprehensive customer intelligence.

- **Marketing:** Provide digital engagement data and implement tracking pixels for cross-platform attribution

- **Sales:** Supply store-level performance data and point-of-sale verification for offline conversion tracking

- **Analytics:** Build data pipelines connecting digital touchpoints to first-party purchase verification

- **Finance:** Establish ROI measurement standards that combine digital spend with verified revenue outcomes
- **Operations:** Ensure inventory and fulfillment data support attribution modeling across channels. Attribution modeling assigns credit for conversions to specific digital touchpoints in the customer journey.

Executive Reflection:

Audit your current digital dashboards and identify which metrics rely on estimated conversions rather than verified purchases. Where could first-party data validation reveal gaps between digital engagement and actual revenue? Which three campaigns from the last quarter would you evaluate differently if you could overlay receipt verification or purchase confirmation data?

Chapter 6 Checklist: Achieving Integrated Insight

- Audit current digital dashboards: Which rely on estimates, not verified outcomes?
- Identify where first-party data can be overlaid (receipts, surveys, loyalty).
- Set up data pipelines to connect digital and first-party datasets.
- Compare results: Which digital channels are correlated with verified purchases?
- Reallocate spend: Increase investment in channels proven to convert.
- Share integrated insights across departments (Strategy, Sales, Product).

Integration clarified which channels generated revenue but also revealed a new challenge: Northbridge's first-party data collection was still ad hoc and inconsistent. Some touchpoints gathered rich customer insights, while others missed opportunities entirely. To scale their competitive advantage, they needed to systematically design every customer interaction as a data collection moment, from initial awareness through repeat purchase, ensuring no valuable signal was lost.

The team's next step was transformative: mapping the 'collection moments' that would fuel their insights engine.

Chapter Seven

Consumer Insight Collection:

Structuring First-Party Data for Growth

Systematic first-party data collection transforms random customer touchpoints into strategic opportunities for gathering intelligence. This chapter provides frameworks for designing collection moments—specifically tailored customer touchpoints that capture first-party data in exchange for clear value—that feel rewarding to customers while capturing decision-grade data that drives business growth.

We'll examine how to structure progressive profiling, gradually collecting customer information over multiple interactions rather than requesting everything at once, which builds trust over time, introduce the Core Data Categories matrix for organizing collection priorities, and provide the Value Exchange framework for ensuring every data request delivers clear customer value.

By the end of this chapter, you'll have practical blueprints for turning every customer interaction into a purposeful data collection moment that strengthens relationships while building competitive intelligence.

Designing Collection Moments

Over the next several weeks, while Finance finalized incentive budgets and Product Development tested seasonal SKUs in pilot markets, the Northbridge marketing team mapped every point where they interact with consumers. Evelyn emphasized that merging digital analytics with first-party data brought new clarity and insight. She coached the group to think from their customers' perspective: every data exchange must feel rewarding and effortless. Simply put, people won't share information without a good reason. The goal is to give them one.

This customer-first perspective directly shaped how each "collection moment" was crafted, those instances when Northbridge invited customers to share first-party data naturally and with obvious value.

At retail displays, the team seizes the chance to capture shopper interest at the point of sale. Northbridge places vibrant signs near beverage coolers and snack aisles, urging: *"Scan to get your $1 off coupon now!" Shoppers who scan are instantly taken to a quick rewards club sign-up—just* enter your email for an immediate discount. This taps immediate purchase intent. A customer already reaching for a drink is prompted: scan to save now and tell us what caught your eye. The result? The shopper wins savings, and Northbridge gains a new contact and real-time product interest. The call to action is clear and compelling, prompting fast engagement without feeling intrusive.

These collection principles also extend to Northbridge's retail and distribution partners. Through co-branded QR kits and signage, partners can host the capture moment while Northbridge still receives the data in real-time, tagged with the partner's store or campaign ID. For example, during a summer launch with a regional grocery chain, each case stack featured a joint Northbridge, retailer sign offering an instant coupon for scanning. The scan counts and conversions appeared instantly in Northbridge's dashboard, segmented by store. This not only proved the display's effectiveness to the retailer but also strengthened the partnership by demonstrating measurable shopper engagement.

Post-purchase moments offer another rich opportunity. Evelyn points out that after someone buys a Northbridge product, they're in a positive frame of mind, they've invested in the brand, and are primed to deepen the relationship. The team designs a seamless follow-up: on the product packaging and the website's thank-you page, a message invites customers to upload their receipt or enter a code for a small reward. For instance, a shopper who just purchased a Northbridge snack pack could snap a photo of the receipt and submit it online to receive bonus loyalty points or a chance at a *"$5 rebate on your next purchase."* This not only delights the customer with a little something extra, but it also gives Northbridge invaluable data (confirming the exact item bought, the store, time of purchase, etc.). By verifying transactions, the marketing team can link the purchase to the individual's profile, thereby bridging offline sales with their digital identity. The consumer, meanwhile, feels that the brand cares about their purchase satisfaction and rewards their loyalty.

In some campaigns, partners even facilitate the post-purchase capture by including a Northbridge QR code (See Appendix E: QR Code Placement Playbook) on their own store receipts or loyalty program emails. This way, the transaction data flows back to Northbridge's unified profile, always retaining the partner's source attribution for tracking purposes.

This process must be quick and user-friendly, such as a simple mobile form or emailing the receipt, so it doesn't feel like a chore. When executed right, many customers will gladly spend a minute for a rebate or points, and in doing so, they volunteer data that marketers usually struggle to get from retailers.

In lifecycle messaging, Northbridge utilizes what Evelyn calls "micro-engagements" to progressively build out customer profiles. Rather than bombarding new subscribers with lengthy surveys about their preferences, the team gradually introduces one-question polls or fun quizzes within emails and social media notifications over time. For example, a week after a customer signs up, they might receive a friendly email newsletter that includes a single, straightforward question: *"Salty or Sweet, what's your 3 PM snack preference?"* with two clickable options. With just one click, the customer has shared a key preference that is automatically added to their profile (sweet-tooth vs. savory-snacker). It doesn't feel like a survey; it feels like a natural part of the content, perhaps followed by a tip like "Thanks! Here's a recipe for a sweet snack mix you might enjoy."

Since answering takes only a second, there's no friction. Over subsequent touch points, the team can ask another casual question, one at a time, such as *"Which Northbridge drink flavor should we launch next?"* or *"Do you pack snacks for work or buy on the go?"* Each interaction is quick, engaging, and provides a nugget of insight. Customers are far more likely to respond to a single, light-hearted question than to a lengthy form. These micro interactions both enrich the data and make the customer feel heard and involved. This *dialogue approach* often boosts engagement rates because consumers enjoy sharing their opinions when it's easy and fun.

Throughout all these collection moments, Evelyn emphasizes progressive profiling. *"Don't ask for everything at once; earn it step by step."* Progressive profiling involves the team gradually gathering data over time, offering something in return each time, thereby motivating the customer to continue. Early on, Northbridge only asks for the basics, say, a contact preference (email, social media, text) and name for the initial sign-up, and makes sure even that comes with a clear benefit (like the instant coupon). As trust grows, later interactions can gently request a bit more, perhaps a birth month (to send a birthday treat) or specific flavor preferences, but always paired with a perk (such as *"Tell us your favorite flavor and get a personalized recommendation!"*).

This way, customers never feel like they're divulging information into a void; there's always an immediate payoff or meaningful response from the brand. Importantly, each data request stays proportionate to the relationship; you wouldn't, for example, ask a first-time customer for their entire demographic profile or income level. Most consumers will gladly provide an email address in exchange for a perk. Still, only a small fraction would feel comfortable sharing more sensitive personal information, such as their household income. That underscores how critical it is to start with low-hurdle questions and save the more personal stuff for later, if at all. Evelyn reminds the team that every new piece of data must feel 100% worth it to the customer.

This could be monetary value (such as discounts or loyalty points) or experiential value (such as personalized content or VIP access). The moment a customer wonders, "Why do they need this info?" without seeing a benefit, the exchange has failed.

To ensure success, the Northbridge team deliberately builds a strong value exchange at each step. They ask: What incentive will truly motivate action? For instance, they know Northbridge snack buyers love early access to new flavors, so one idea is a *"Taste Testers Club."* Customers who share a couple of extra profile details (like their flavor preferences and mailing address) could be invited to get free samples of upcoming products before the general public. The message is direct: Share more, get exclusive treats—act now! This makes customers *eager* to provide more data in exchange for a tangible, exciting reward.

Another example: the team gamifies data collection with a "Snack Personality Quiz" on the Northbridge website or social media. Questions like *"Pick a movie night snack spread"* or *"What's your go-to energy boost?" culminate in a prompt: "Enter your email now to see your snack profile and get a special recipe!" The consumer is entertained and instantly rewarded, while Northbridge collects valuable insights. By making the ask clear and irresistible, the team transforms data collection into an engaging invitation to act.*

Sidebar: Real-world example

Promolytics gives you the tools to enable partners to collect marketing data on your behalf, seamlessly merging it into individual consumer profiles while preserving source attribution. This enables you to measure channel-level ROI without compromising compliance.

Evelyn underscores that trust and respect are paramount in motivating consumers. The team ensures that every request is transparent and contextually relevant. If Northbridge is requesting a piece of information, it's either obvious why (asking for a zip code to find the nearest store, for example) or explained upfront ("Share your birthday and we'll send a special birthday reward!").

This transparency, combined with consistent delivery of promised value, builds goodwill. Over time, as customers experience the benefits – including applicable coupons, engaging content, and improved product recommendations – they become more willing to share and participate.

In essence, Northbridge is nurturing a two-way relationship. Rather than the old-school approach of simply *taking* data, the team focuses on *earning* data. They treat each collection moment as a mini customer engagement campaign in itself.

Finally, Evelyn reminds the team that consumers generally show greater willingness to share information when they receive clear value in return, such as loyalty rewards or exclusive perks. This is proof that the marketing team's efforts to provide clear rewards will directly pay off in higher participation. Consumers respond to meaningful rewards, whether monetary savings or exclusive experiences, and some will even switch brand loyalties for a better value proposition.

Armed with this understanding, the team commits to designing every data touchpoint as a motivating, value-packed moment. Together, they map the consumer journey, marking exactly where and how to prompt action—whether to scan, sign up, upload, or answer a poll. Every step should prompt the customer: "Take this action, get this reward—now!" With this approach, each collection moment builds engagement and leaves consumers eager for the next opportunity to interact.

From Moments to Master Blueprint

In **Designing Collection Moments,** you learned *how* to motivate consumers to share the correct information at the right time. Now we need to decide *what* information truly matters, *why* it matters, and *how* each piece will be integrated into the business once captured. Consider the following table, **Core Data Categories**

and Use Cases, as the master blueprint that consolidates scattered touchpoints into an integrated data engine.

1.Scope & Structure

- Each row represents an authorized data category (e.g., Identity, Behavioral, Transaction).
- For every category, the table maps collection sources, exemplary fields, and the business questions those fields unlock.
- It also flags which stage of the **INSIGHT Framework** (Chapter 9) the data feeds belong to, making it clear where each column comes to life: Identify, Normalize, Segment, Interpret, Guide, Harness, or Track.

2. Governance & Accountability

- By explicitly tying every data point to a *primary KPI* and its corresponding *qualitative impact*, you ensure that teams collect data with purpose, not out of curiosity.
- The table becomes the single source of truth for legal, analytics, marketing, and product teams, eliminating the confusion that often derails data projects, such as "what does this column even do?"

3. Seamless Progression to Value Exchange

- Once you know *which* data matters and *when* you're asking for it (as outlined in **Designing Collection Moments**), the **Value Exchange Matrix** will help you craft the '*why*' that motivates the consumer, linking every data request to a clear benefit and incentive.
- Together, the three elements form a pipeline:

1. **Moment** → Where/When we ask
2. **Category** → What we ask for and why it's valuable internally
3. **Exchange** → How we reward the consumer for providing it

As you review the **Core Data Categories and Use Cases** table, picture it hanging on the wall in the board room: every new campaign, survey, or rebate program should trace its data demands back to this blueprint. Anything that doesn't map (or can't articulate a KPI) gets cut. That discipline is what turns first-party data collection from a scatter-shot tactic into a scalable growth asset.

Core Data Categories and Use Cases

The working group formalizes a definitive **Core Data Category** table. This structure becomes the blueprint for governance, analytics, and activation.

Let's pause here: Up to this point, you've sketched where data is captured through carefully designed collection moments. The next logical question is, "Which specific data points move the needle once they land in your system?" The table that follows is Northbridge's master ledger, the definitive list of every data category the team will govern, analyze, and activate. Scan it first for familiar staples, such as identity or Transaction data, and then note newer categories, like sentiment and operational execution, that often hide untapped value. Use the rightmost columns as your roadmap: each KPI tells you precisely how success will be measured and where in the INSIGHT Framework that data will ultimately create lift.

To operationalize the table, assign an owner to each category and publish a short data dictionary that lists fields, permissible uses, consent basis, refresh cadence, and retention rules. Set QA checks for accuracy, relevance, and freshness, and schedule a monthly review to retire fields that do not tie to a KPI or a live use case.

Require every campaign brief to cite the specific rows it will use and the INSIGHT stage it supports, so budgets fund data that moves outcomes. Post the table in the boardroom and in your analytics workspace; it becomes the reference that keeps Marketing, Sales, Product, Finance, and Legal aligned, making every new 'ask' feel purposeful to the customer and valuable to the business.

Category	What It Is	Exemplary Fields / Elements	Primary Collections Methods (Northbridge Example)	Key Decisions Enabled (Qualitative Impact)	INSIGHT Mapping	Primary KPIs (Examples)
Identity & Profile	Stable attributes about an individual or account	Name, email, role, geography, device ID, consent status	Account forms, QR opt-in, loyalty enrollment	Audience sizing, permissioned outreach cadence	Identify, Normalize	Net new verified contacts; percent with valid consent; profile completeness rate
Behavioral (Digital)	On-site or in-app actions	Page views, feature clicks, session depth	Cookies, tagged links, product analytics	Funnel refinement, feature prioritization	Identify, Segment, Interpret	Activation rate; time-to-first value; adoption depth
Behavioral (Traditional Promotion)	Offline interactions signals tied to a marketing promotion	Event attendance, sampling acceptance, coupon scans	QR on printed materials at retail, displays, receipt upload, staff tablets, media	Channel Mix allocation, experiential ROI	Identify, Normalize, Track	Event participation rate; sample to purchase confirmation percent; first-party consumer feedback
Transactions & Commerce	Monetary interactions	SKU, units, price paid, basket composition	Receipt validation, POS feed, loyalty code entry	Promotion lift attribution, pricing refinement	Normalize, Track, Guide	Incremental lift percent; repeat purchase interval
Engagement & Communication	Response to outbound messaging	Email open dwell, click intent, unsubscribe reason	ESP logs, preference center	Cadence tuning, fatigue prevention	Segment, Interpret, Guide	Engaged contact ratio; click to conversion rate
Offer & Incentive Response	Reaction to coupons, trials, bundles, experiences, incentives	Offer type, redemption latency, repeat usage	QR coupon generation, unique URL	Incentive efficiency, spend reallocation	Interpret, Track, Guide	Redemption rate; breakage percent
Preference & Intent	Stated wants & timing signals	Planned purchase window, flavor interest	Micro-surveys, progressive profiling	Launch timing, assortment planning	Identify, Segment, Interpret	Preference capture rate; staged to actual conversion percent
Contextual & Environmental	Situation around interaction	Location, device, time of day	Geo-IP, location, timestamps	Daypart optimization, regional allocation	Normalize, Interpret, Track	Daypart performance variance; geo-conversion differential
Sentiment & Qualitative Feedback	Perception & emotion drivers	NPS verbatims, satisfactions reasons	Post experience QR survey, CSAT	Messaging reframes, product backlog scoring	Interpret, Guide, Harness	NPS; sentiment shift
Lifecycle & Cohort State	Stage in customer journey	New, engaged, lapse risk, advocate	Journey analytics, retention tagging	Churn prevention triggers, expansion timing	Segment, Guide, Harness	Cohort retention percent; re-engagement success rate
Predictive / Scored Attributes	Propensities derived via modeling	Likelihood to repurchase, churn probability	Unified warehouse modeling	Proactive outreach, forecast accuracy	Interpret, Guide, Harness	Model precision
Cost to Serve & Efficiency	Resource utilization by segment	Support hours per cohort, fulfillment cost	Support system ingestion	Margin mix optimization	Track, Guide	Support minutes per revenue dollar
Consent & Compliance Signals	Permissions & lawful basis	Opt-in source, revocation timestamp	Consent management service	Lawful personalization boundaries	Identify, Normalize, Track	Consent opt-in rate; consent decay rate
Cross Category Interaction	Multi-line product relationships	Category overlap, sequential adoption path	Unified purchase & engagement history	Cross-sell sequencing, bundle design	Segment, Interpret, Guide	Cross-sell rate; attach rate
Operational Execution / Field Data	In-market executions quality	Display compliance, stock-out incidence	Field rep mobile app, photo capture	Remediation of execution gaps	Track, Guide	Display compliance percent, out-of-stock rate

Core Data Categories & Use Cases

While Marketing finalized this table, Sales prepared training for frontline teams to reinforce how to explain value exchanges during in-store activations.

Introducing the Value Exchange Matrix

Having identified your core data categories and the use cases they unlock, the next task is to decide how you will earn each data point from the consumer. The Value Exchange Matrix on the next page translates that challenge into a practical worksheet. For every piece of information you hope to collect (an email address, retailer location, flavor preference, and so on), the matrix forces you to answer five questions:

1. **Internal Business Value** – Why does this data matter to the brand?
2. **Consumer-facing Benefit** – What does the shopper gain in return?
3. **Incentive Structure** – Which reward or experience best delivers that benefit?
4. **Microcopy Example** – Exactly how will you phrase the ask in plain, motivating language?
5. **KPI to Watch** – Which signal tells you the exchange is working or needs refinement?

Use the matrix as a living document. Populate it row by row during campaign planning, attach cost ceilings to each incentive, and compare the expected data value to the reward you offer. If the value falls short, rework the incentive or skip the request entirely. Treated this way, the matrix becomes a guardrail, keeping every data collection moment accountable to both the consumer experience and the bottom line.

Once you complete this exercise, you will be ready to map each exchange to the broader INSIGHT Framework previewed in this chapter.

Value Exchange Matrix

From blueprint to bargain: Once Northbridge knows what data to capture, the next step is negotiating a fair trade with the consumer. Every bit of data is exchanged for value—whether it's a rebate, a recipe booklet, or an Insider Panel ticket. The "Microcopy Example" column is crucial: a single motivating line can determine conversion. If the incentive isn't clear, reconsider the ask.

Data Point	Internal Business Value	Consumer-facing Benefit	Incentive Structure	Microcopy Example	KPI to Watch
Email/Mobile	ID, follow-up offers, analytics	Fast delivery of a rebate or coupon	Instant reward (rebate), nurture offers	"Enter your email to receive your rebate in minutes."	Form completion rate
Retailer / Zip	Channel ROI, geo-targeting	Localized deals, find-it-near-me	Local coupons, store locator	"Tell us where you found it so we can send local specials"	% users providing location
Flavor Preference	Product roadmap input	Personalized content	Recipe booklet, sample giveaway	"Pick what you like, we'll only send the good stuff"	Click-through on flavor-specific emails
Usage Occasion	Campaign targeting	Relevant ideas (party packs, gifting)	Sweepstakes entry tied to the occasion	"How did you enjoy it? One tap answer = entry to win"	Response rate, contest entries
Purchase Frequency	Forecasting, LTV scoring	Loyalty tiers, bulk offers	Tiered rewards (Bronze/Silver/Gold)	"Share how often you buy to unlock higher-value perks"	LTV lift, repeat purchase rate
Demographic Band (age range)	Compliance, segmentation	Age-appropriate content, legal compliance	Access to age-gated content	"Choose your range. We do not need the exact age."	Drop-off at age question
Opt-in to Panels	Research agility	Early access, influence roadmap	Beta invites, exclusive swag	"Want to help shape what's next? Join our Insider Panel"	Panel signup rate
Receipt Image Metadata (date, total)	Fraud checks, basket analysis	Faster reimbursements next time	Auto-approval next rebate	"Upload once. We'll auto-approve faster next time"	Time to reimbursement

Value Exchange Matrix

Tip: Attach each incentive to a cost ceiling and an expected data value score. If value score < incentive cost, do not ask.

Privacy Compliance: All collection moments require transparent consent and a clear explanation of the value. Sample language: 'We use this information to personalize your experience and send relevant offers. You control your preferences and can opt out anytime.' Ensure compliance with applicable privacy laws. See Chapter 11 for complete protocols.

Cross-Functional Implementation

Systematic data collection requires coordinated execution across all customer-facing departments. Each team must understand collection opportunities, value exchange principles, and data quality standards to ensure consistent customer experiences.

- **Marketing:** Design collection moments with clear value propositions and implement progressive profiling across campaigns

- **Sales:** Train promotional teams on value exchange explanations and capture store-level interaction data

- **Customer Service:** Integrate collection opportunities into support interactions and post-resolution follow-ups

- **Product:** Use verified customer feedback for development priorities and feature validation

- **Legal:** Ensure all collection methods comply with consent requirements and privacy regulations

Mapping to the INSIGHT Framework: A Preview

Each data touchpoint in this chapter aligns with our seven-step **INSIGHT Framework**, a closed-loop system for turning first-party data into revenue-driving action. We'll explore the full mechanics, tools, and KPIs for every stage in Chapter 9. For now, keep this high-level flow in mind:

1. **Identify** → capture core identity and consent

2. **Normalize** → unify data feeds under common keys

3. **Segment** → group customers dynamically

4. **Interpret** → derive sentiment and predictive insight

5. **Guide** → translate findings into playbooks

6. **Harness** → trigger automated campaigns

7. **Track** → measure lift and feed learnings back to identify

Think of these seven steps as the backbone of every collection tactic you implement. We'll dive deep in Chapter 9.

Northbridge Illustration: Systematic Collection in Action

Six weeks after implementing their Core Data Categories framework and Value Exchange Matrix, Northbridge launched a coordinated collection strategy across their summer citrus launch. The systematic approach replaced ad hoc data gathering with structured intelligence collection at every customer touchpoint.

Collection Moment Design (Week 1-2):

- **Retail displays:** QR codes offering $1.50 instant rebates captured email + flavor preference
- **Post-purchase:** Receipt upload prompts with 500 bonus loyalty points for transaction verification
- **Email follow-up:** Single-question polls embedded in newsletters ("Citrus intensity: Mild or Bold?")
- **Social media:** "Snack Personality Quiz" requiring contact information (email, social media, SMS) for results delivery

Progressive Profiling Implementation:

- **Phase 1 (Immediate):** Email + purchase intent (95% completion rate)
- **Phase 2 (7 days later):** Flavor preference via one-click poll (73% response rate)
- **Phase 3 (14 days later):** Usage occasion through gamified quiz (61% participation)
- **Phase 4 (30 days later):** Purchase frequency for tiered rewards (47%

engagement)

Data Quality Metrics (45 days):

- **Accuracy rate:** 97% (verified against POS data)
- **Consent compliance:** 100% (clear opt-in at every touchpoint)
- **Profile completeness:** Average 6.3 data points per customer (up from 1.8)
- **Data freshness:** 94% of records updated within 30 days

Business Impact by Category:

- **Identity & Profile:** 4,847 new verified customer profiles with complete contact preferences
- **Behavioral Insights:** 78% preference correlation with actual purchase patterns
- **Transaction Verification:** $127K incremental revenue attributed to collection-driven targeting
- **Engagement Patterns:** 41% improvement in email open rates through preference-based segmentation

Cross-Functional Activation:

- **Marketing:** Created 12 dynamic segments based on collected preferences, achieving 2.3x higher conversion rates

- **Sales:** Used verified regional preferences to secure 34% more shelf space in high-citrus-preference markets
- **Product:** Prioritized "Bold Citrus" line extension based on 67% preference for intense flavors
- **Operations:** Adjusted inventory allocation, reducing stockouts by 28% in preference-matched regions

Value Exchange ROI:

- **Average incentive cost:** $0.73 per complete profile
- **Customer lifetime value increase:** $14.60 per systematically profiled customer
- **Net value creation:** $13.87 per customer relationship
- **Program ROI:** 1,897% return on collection investment

Governance Success Indicators:

- **Data stewardship:** Zero compliance issues across 15 collection touchpoints
- **Quality maintenance:** Monthly review process identified and resolved 3% data discrepancies
- **Consent management:** 94% customer satisfaction with transparency and value delivery

Compound Effect (90 days): Customers acquired through systematic collection demonstrated 3.4 times higher repeat purchase rates and 2.1 times greater email engagement compared to traditional acquisition methods. The structured approach enabled Northbridge to predict seasonal demand with 89% accuracy, reduce customer acquisition costs by 31%, and improve data quality across all touchpoints.

Key Insight: Systematic collection of moments generated 847% more actionable customer intelligence per interaction compared to ad hoc data gathering, while maintaining 94% customer satisfaction with the data exchange process.

This transformation demonstrated that structured collection creates compounding value; each interaction builds upon previous insights to deliver increasingly precise customer understanding and business results.

Sidebar: Real-world example

Platforms such as Promolytics operationalize phased capture using QR scans, URLs, surveys, cookies, and receipt validation to enrich records over time. By integrating each step, from first scan to post-purchase receipt, into a unified customer dashboard. Promolytics ensures every data point is tied to a verified, unique consumer profile.

For example, a beverage brand could run a "Flavor of the Month" campaign using QR codes (See Appendix E: QR Code Statistics & Placement Playbook) on in-store displays and packaging. Scanning leads to a one-question poll, followed by a rebate offer upon receipt upload. As data flows in, each unique customer profile is created or updated depending on whether it is their first interaction with the brand. The platform then enables the team to segment participants by any data point collected and stored in the customer's profile, such as flavor preference, purchase location, and behavior. Marketing utilizes these segments to retarget high-value groups with tailored offers, while Sales leverages the verified results to negotiate more favorable shelf placement and displays. This process closes the loop from engagement to activation, all without relying on third-party data.

Executive Reflection – Foundation Readiness:

Map your organization's current customer touchpoints and identify which interactions currently capture first-party data versus which represent missed opportunities. Using the Core Data Categories framework, which three data types would most improve your decision-making if captured consistently? Design one collection moment that provides clear customer value while gathering strategically important intelligence within the next 30 days.

Chapter 7 Checklist: Building the First Party Foundation

- Inventory current data against the category table and mark gaps.
- Implement progressive profiling for at least one high-traffic touchpoint.
- Establish a unique identity key for each consumer and apply it to all collection assets.
- Assign stewards and schedule a monthly data quality review.
- Launch one activation use case using a newly defined dynamic segment.

Northbridge had systematically designed collection moments that captured high-quality first-party data through clear value exchanges. But collection alone doesn't create competitive advantage—the real power emerges when those scattered signals become a self-reinforcing system that generates smarter decisions with each cycle.

The next challenge was building the engine that would transform individual data points into continuous intelligence: the Consumer Insights Flywheel that turns today's customer signals into tomorrow's competitive advantages.

Chapter Eight

Building the Consumer Insights Flywheel:

When Information Generates Self-Reinforcing Growth

The Consumer Insights Flywheel transforms scattered first-party data into a self-reinforcing competitive advantage. This chapter reveals how to design a closed-loop system where each customer interaction fuels analysis, enables smarter activation, and generates richer signals that accelerate learning with every cycle.

We'll examine the six stages that convert individual data points into continuous intelligence, introduce acceleration levers that add momentum to each rotation, and provide frameworks for partner-powered data collection that scales insights without proportional costs.

By the end of this chapter, you'll have blueprints for building flywheel systems that compound customer understanding, operational frameworks for sustaining momentum across departments, and measurement protocols that ensure each cycle delivers measurable business acceleration.

Northbridge: Turning Insight Into Momentum

Midway through the quarter, while Marketing finalized assets for upcoming regional promotions and Data Engineering prepared the next CRM integration release, Evelyn convened a meeting with leaders from Marketing, Sales, Customer Service, Product Development, and Strategy. On a whiteboard, she sketches a circle with six arrows feeding clockwise.

"This is our Consumer Insights Flywheel," she says, "Keep it turning and learning accelerates, so we outpace rivals with precise, personalized content."

The team leans in. Over the past few months, they have addressed blind spots, reduced unverified spending, and established a foundation for first-party consumer insights. Now they will convert insight into accelerating momentum.

What a Flywheel Does

A flywheel stores kinetic energy through rotational momentum. Each spin adds force, making subsequent cycles easier and more powerful. In business terms, every customer interaction collected fuels deeper analysis, which leads to improved activation strategies, yielding richer interactions and exponentially more valuable data—the system compounds upon itself. *Figure 1* visualizes this self-reinforcing, closed-loop engine.

The beauty of the flywheel model lies in its sustainability and acceleration over time. Unlike traditional linear growth models that require constant input to maintain momentum, a well-designed flywheel becomes increasingly efficient as it gains speed. Initial effort may feel substantial, but as each component strengthens the others—better data enabling smarter decisions, smarter decisions creating superior customer experiences, and superior experiences generating more valuable data—the system begins to power itself forward with minimal additional energy input.

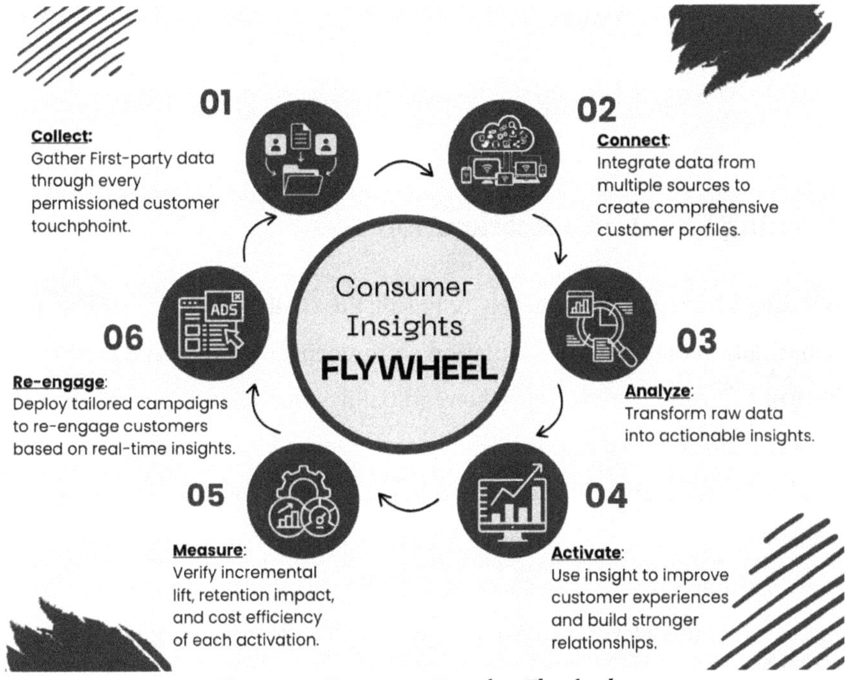

Figure 1: Consumer Insights Flywheel

The Six Stages

1. **Collect** – Capture permissioned, context-rich signals at every touchpoint: QR scans, URL links, consumer insights, receipt uploads, web clicks, support calls.

2. **Connect** – Resolve identities and standardize events so browsing, purchase, and feedback converge on a single profile.

3. **Analyze** – Segment, score, and interpret patterns that explain behaviors and predict future outcomes.

4. **Activate** – Deliver personalized offers, content, or product changes based on real-time first-party insights.

5. **Measure** – Verify incremental lift, retention impact, and cost efficiency

of each activation.

6. **Re-engage** – Loop results back into the process, then iterate.

Turning Speed into Acceleration

Building a functional flywheel is only half the challenge; the other half is systematically accelerating each rotation so momentum compounds rather than plateaus. Organizations often achieve initial flywheel success but fail to sustain growth because they don't deliberately engineer acceleration into each cycle.

The six acceleration levers below transform steady flywheel rotation into exponential momentum. Each lever addresses a specific friction point that typically slows learning velocity or reduces data quality. Rather than hoping for organic acceleration, these levers provide systematic methods for adding torque to every stage of the insights cycle.

Lever	Description	Northbridge Action Example
Value Exchange	Strong incentives that motivate data sharing	Instant rebate for receipt upload
Data Unification	Rapid integration of new fields into the profile	First-party and Digital data integrated for a comprehensive view
Analytical Maturity	Use of predictive scores and casual testing	Churn propensity score guiding offers
Activation Speed	Automated journeys are triggered within minutes	Receipt upload fires thank-you email and cross-sell coupon
Measurement Discipline	Holdout groups and lift reports are reviewed weekly	Capture Pulse dashboard flags underperforming variant
Governance Alignment	Consent, quality, and security checks embedded	Unified consent ledger gates outbound audience exports.

Acceleration Levers

Implementing acceleration levers requires disciplined experimentation rather than wholesale changes. Select one lever per flywheel cycle, assign clear ownership with measurable targets, and run controlled tests with 14-21 day evaluation windows. Document which levers generate the most torque for your specific business context—some organizations see dramatic acceleration from improved value exchange, while others benefit most from analytical maturity or governance alignment.

The goal isn't to deploy all levers simultaneously, but to systematically identify and eliminate the highest-impact friction points that prevent your flywheel from reaching optimal velocity. Each successful lever implementation should increase learning velocity while maintaining data quality and customer satisfaction, creating sustainable momentum that compounds over multiple cycles.

Northbridge Illustration : Flywheel in Full Rotation

When Northbridge launched its "Citrus Burst" limited-edition flavor, it deployed its Consumer Insights Flywheel for the first time as a complete system. The six-week campaign demonstrated how each stage accelerated the next, creating compounding momentum that transformed a product launch into a learning engine.

Week 1-2: Collect Stage

- **Retail displays:** 847 QR scans offering exclusive early access captured email + flavor intensity preference
- **Social media:** "Flavor Quiz" generated 1,234 email captures with usage occasion data
- **Partner events:** Regional grocery chain sampling events added 456 verified profiles
- **Post-purchase:** Receipt uploads confirmed 312 actual purchases with transaction details
- **Collection efficiency:** 94% data quality rate, 97% consent compliance

Week 2-3: Connect Stage

- **Identity resolution:** Unified 2,134 touchpoints into 1,847 unique customer profiles (87% match rate)
- **Data integration:** Combined digital engagement, purchase behavior, and stated preferences within 24 hours
- **Profile enrichment:** Average customer record increased from 2.1 to 5.7 data points
- **Technical performance:** Real-time data ingestion reduced lag from 48 hours to 6 minutes

Week 3-4: Analyze Stage

- **Segmentation:** Identified "Intense Citrus" segment (34% of respondents) with 3.2x higher purchase intent
- **Predictive scoring:** Churn propensity model flagged 127 at-risk customers for retention campaigns
- **Regional patterns:** Northeast markets showed 67% preference for "Bold" intensity vs. 23% in Southeast markets
- **Behavioral insights:** "Workplace snackers" segment demonstrated 2.4x higher lifetime value potential

Week 4-5: Activate Stage

- **Targeted campaigns:** Personalized emails to "Intense Citrus" segment achieved 47% open rates (vs. 23% baseline)

- **Regional customization:** Adjusted creative messaging by intensity preference, improving CTR by 156%
- **Inventory optimization:** Shifted 40% more "Bold" variants to Northeast markets based on preference data
- **Partner activation:** Shared insights with retail partners to optimize shelf placement and promotional timing

Week 5-6: Measure Stage

- **Conversion lift:** Targeted segments converted 3.7x higher than control groups
- **Revenue attribution:** $247K incremental sales directly traced to flywheel-driven personalization
- **Retention impact:** Personalized customers showed 89% satisfaction vs. 67% for generic campaigns
- **Efficiency gains:** Cost per acquisition decreased 31% through precision targeting

Week 6+: Re-engage Stage

- **Learning integration:** Insights fed into "Tropical Blend" launch planning, predicting optimal flavor intensity
- **Customer deepening:** Follow-up surveys captured usage occasion data from 73% of engaged customers
- **Partner expansion:** Success metrics convinced 3 additional retail chains to adopt the co-branded collection

- **Continuous optimization:** Weekly micro-tests refined messaging, improving response rates by 12% monthly

Flywheel Acceleration Metrics:

- **Learning velocity:** Insight-to-activation time reduced from 3 weeks to 4 days
- **Data quality improvement:** 97% accuracy rate (up from 78% pre-flywheel)
- **Customer engagement lift:** 234% increase in voluntary data sharing over 6 weeks
- **Cross-functional adoption:** All 5 departments are actively using Flywheel insights for decision-making

Partner-Powered Amplification: The regional grocery partner's checkout QR codes contributed 1,156 additional profiles, while Northbridge's co-branded festival booth captured 623 new customers. Partner attribution revealed that 43% of their customers also engaged with Northbridge's direct touchpoints, uncovering previously invisible customer journey patterns.

Privacy Compliance: Partner data collection requires clear consent attribution and shared responsibility protocols. Sample language: 'Data collected through our partners helps us improve your experience across all touchpoints.' Ensure all partner agreements include privacy compliance requirements. See Chapter 11 for complete protocols.

Compound Effect (90 days post-launch): The "Citrus Burst" flywheel generated insights that improved three subsequent product launches. Customer profiles built during the initial campaign achieved 4.2 times higher engagement rates and 67% lower churn rates compared to traditionally acquired customers. The flywheel's momentum enabled Northbridge to predict seasonal demand fluctuations with 91% accuracy while reducing customer acquisition costs by 48%.

Key Breakthrough: Week 4 analysis revealed that customers who engaged across multiple touchpoints (QR scan + social quiz + receipt upload) showed a 6.8 times higher lifetime value. This insight transformed Northbridge's collection strategy from single-touchpoint optimization to multi-channel engagement orchestration.

The campaign proved that flywheel momentum compounds: each rotation taught the system to identify higher-value customers faster, activate them with greater precision, and measure results with increased accuracy, creating sustainable competitive advantages that strengthened with every customer interaction.

Sidebar: Real-world example

Promolytics provides brands with tools to keep their consumer insights flywheel spinning, uploading SMS, QR capture, surveys, receipt validation, and activation triggers into unique customer profiles—even when data comes from retail or event partners.

With this unified view, marketers can quickly launch targeted offers, sales teams can track lift in real-time, and product teams receive immediate feedback. Feeding verified results back into campaign design accelerates the cycle for smarter, more profitable activations.

Cross-Functional Implementation

Sustaining Flywheel momentum—the self-reinforcing acceleration that occurs when each customer insight generates better actions, which in turn create richer customer interactions—requires coordinated execution across all departments. Each team contributes specific inputs and takes ownership of particular stages to ensure continuous acceleration.

- **Marketing:** Own collection and activation stages, design progressive profiling campaigns, measure engagement lift
- **Analytics:** Manage, connect, and analyze stages, ensure data quality, and build predictive models
- **Sales:** Provide retail performance data, use insights for shelf placement negotiations, and validate store-level lift
- **Product:** Interpret customer feedback signals, prioritize feature development based on verified demand
- **Customer Service:** Capture support interaction data, identify friction points, measure satisfaction improvements

Action Steps Implemented

Each rotation revealed new insights and exposed small friction points that could be smoothed in the next pass. To capture and reinforce these learnings, the team documented concrete action steps, turning the Flywheel from concept into a daily operating practice.

- Added real-time ingestion of QR scans.
- Configured real-time identity resolution to reduce duplicate contact records.
- Deployed an automated nurture sequence triggered by User Experience Submission.
- Implemented a Flywheel Review process: conducted an interim analysis one week after launch to adjust tactics as needed, followed by a post-campaign review to capture learnings and optimize future initiatives.

Executive Reflection:

Map your organization's current data-to-activation cycle and identify where the most friction occurs between collecting customer signals and taking action on insights. Which of the six flywheel stages represents your biggest bottleneck, and what single improvement would most accelerate learning velocity—the speed at which customer insights translate into improved business decisions and outcomes? Design one test using partner-powered data collection that could add momentum to your insights flywheel within 30 days.

Chapter 8 Checklist: Spinning the Flywheel Faster

- Establish a cross-functional Flywheel Review.
- Create a new capture point and confirm that data updates into the unique customer profile.
- Set up a basic holdout control for any new activation to measure lift.
- Select and test one acceleration lever from the framework to improve flywheel velocity.
- Document and iterate on learning velocity (days from insight to deployment).

Northbridge's flywheel spun faster with each campaign, building momentum that accelerated learning and boosted results. However, sustaining this pace required more than intuition; it demanded systematic operating procedures with clear ownership, defined measurement standards, and reliable stage handoffs.

The solution was INSIGHT: a seven-stage framework— Identify, Normalize, Segment, Interpret, Guide, Harness, Track—that transformed the flywheel from an inspiring concept into a disciplined operating system with defined roles, measurable outcomes, and scalable processes.

Chapter Nine

The INSIGHT Framework:

When Seven Stages Transform Data Into Decisions

The INSIGHT Framework transforms first-party data management from ad hoc activities into disciplined operations with clear ownership, measurable outcomes, and systematic handoffs. This chapter presents a seven-stage operating system that converts customer signals into verified business results through coordinated, cross-functional execution.

We'll examine the core questions and deliverables of each stage, introduce ownership models that prevent bottlenecks, and provide cadence frameworks that ensure consistent execution velocity—the speed at which approved strategic decisions are implemented and deployed across customer touchpoints. You'll learn to establish measurement protocols that track learning velocity—the speed at which customer insights translate into improved business decisions and outcomes—and business impact across the complete insights lifecycle.

By the end of this chapter, you'll have operational blueprints for implementing systematic first-party data processes, governance frameworks that maintain quality and compliance, and measurement systems that prove ROI while accelerating continuous improvement.

Northbridge: INSIGHT in Practice

Later in the quarter, as the analytics team finalized campaign reports and sales prepared for line reviews, Evelyn convened the leadership team to introduce their new operational framework. On the whiteboard, she wrote seven letters: I-N-S-I-G-H-T.

"Each letter represents a stage we must execute with discipline," she explained. "Together they form the INSIGHT Framework—our operating system for first-party data that will align every department around systematic customer intelligence."

The INSIGHT Framework serves as a blueprint for transforming first-party data into actionable results. It organizes work into seven stages that mirror value creation, from capturing permissioned signals to proving revenue impact. The framework aligns marketing, sales, product, and data teams around shared language, deliverables, and cadence while ensuring each stage produces tangible achievements: better signal identification, cleaner data, sharper customer segments, clearer interpretation, informed decisions, effective activation, and measurable business impact.

Each stage generates a concrete artifact that advances the process: a capture-point map (**Identify**) clarifies entry points, a data model with governance checklist (**Normalize**) ensures quality standards, a segmentation charter (**Segment**) defines target groups, an insight brief (**Interpret**) highlights hypotheses, a decision log (**Guide**) records choices, an activation plan (**Harness**) summarizes execution steps, and a measurement plan (**Track**) captures ROI and learning progress. Regular artifact reviews and updates build organizational momentum each quarter. (Figure 2)

The INSIGHT Framework

A practical framework to help organizations operationalize their first-party data strategies across departments.

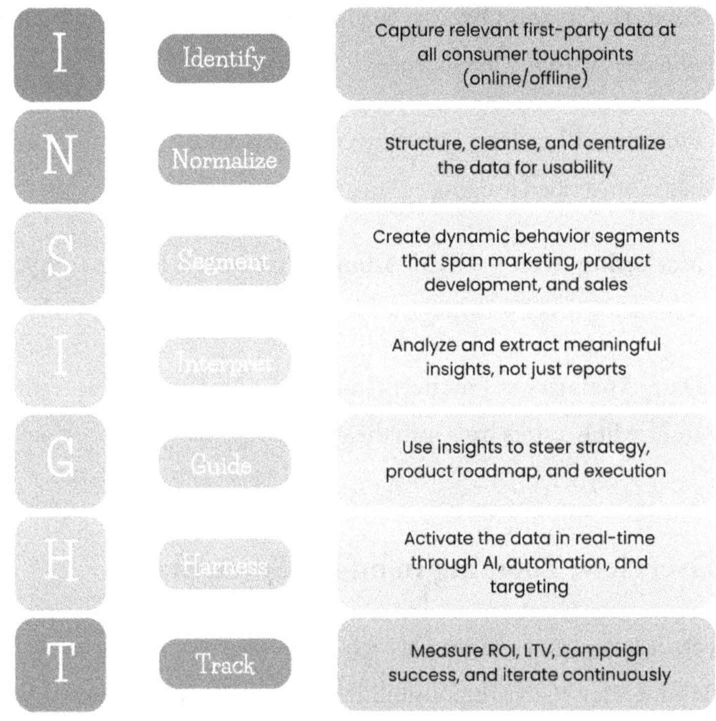

I
Identify
Capture relevant first-party data at all consumer touchpoints (online/offline)
N
Normalize
Structure, cleanse, and centralize the data for usability
S
Segment
Create dynamic behavior segments that span marketing, product development, and sales
I
Analyze and extract meaningful insights, not just reports
G
Guide
Use insights to steer strategy, product roadmap, and execution
H
Harness
Activate the data in real-time through AI, automation, and targeting
T
Track
Measure ROI, LTV, campaign success, and iterate continuously

Figure 2: The INSIGHT Framework

Cross-Functional Implementation

The INSIGHT Framework succeeds only when every department understands its role in the system and commits to reliable handoffs between stages. Each functional team owns specific stages while contributing to others, creating accountability without silos.

- **Marketing Ops + Data Engineering:** Co-own Identify stage, establishing capture points and consent protocols
- **Data Engineering:** Lead the Normalize stage, ensuring data quality and integration standards
- **Analytics Team:** Own the Segment and Interpret stages, building actionable customer intelligence
- **Functional Leaders:** Drive the Guide stage, making strategic decisions based on verified insights
- **Marketing Tech + CRM Admin:** Execute the Harness stage, operationalizing insights through automation
- **Data Analytics + Finance:** Co-own the Track stage, measuring ROI and feeding results back into the system

Stage Overview: Zooming in on the Seven Stages

The flip-chart letters gave the team a high-level view; the table below translated those letters into day-to-day responsibilities. Each column functioned like a mini job description: "What question are we answering?", "What activities make that answer possible?", "Who owns the outcome?" This operational clarity prevents the framework from becoming theoretical; every stage has measurable deliverables and accountable owners. Any cell lacking clarity was a signal to act; mis-ownership is the fastest way to stall an otherwise powerful framework.

Stage	Core Question	Primary Activities	Ownership Anchor
Identify	What customer signals can we capture today.	Define capture points, consent language, identity keys.	Marketing Ops + Data Engineering
Normalize	How do we make disparate data usable.	Clean, deduplicate, standardize time zones, apply metadata.	Data Engineering
Segment	Which groups exhibit shared behaviors or value.	Dynamic grouping by lifecycle, propensity, offer response.	Analytics Team
Interpret	Why do segments behave as they do, and what will they do next.	Root-cause analysis, predictive modeling, qualitative overlay.	Insights Lead
Guide	Which business decisions must change.	Budget reallocation, product backlog reprioritization, sales focus.	Functional Leaders
Harness	How do we operationalize insights in real time.	Triggers, personalization rules, automated workflows.	Marketing Tech + CRM Admin
Track	Did the action create incremental value.	Holdout tests, KPI dashboards, learning velocity metrics.	Data Analytics + Finance

INSIGHT Activities and Responsibilities

Northbridge Walkthrough: INSIGHT Framework in Action

When Northbridge launched their "Berry Fusion" limited edition, they deployed the complete INSIGHT Framework for the first time on a major product introduction. The 21-day campaign demonstrated how systematic stage execution with defined handoffs accelerated decision-making and measurable business impact.

Day 1-3: Identify Stage (Marketing Ops + Data Engineering)

- **Deliverable:** Capture point map with 12 touchpoints across retail, digital, and partner channels
- **Execution:** QR codes on 1,847 shelf displays offering early access captured preferred communication method (email, social media, SMS) + flavor intensity preference

- **Quality metrics:** 96% data capture rate, 100% consent compliance, identity key assignment for new consumers in database
- **Handoff criteria:** Verified data structure, consent documentation complete, unique ID assignment functional
- **Stage owner accountability:** Marketing Ops confirmed 2,156 permissioned profiles ready for normalization

Day 4-6: Normalize Stage (Data Engineering)

- **Deliverable:** Unified customer database with standardized schemas and quality scorecards
- **Execution:** Real-time ETL (Extract, Transform, Load) process—a data integration methodology that extracts information from multiple sources, transforms it into standardized formats, and loads it into a unified database—ingested survey and receipt data, updating existing profiles or creating new unique IDs within 15 minutes.
- **Quality metrics:** 94% identity match rate, 99.2% data accuracy against validation rules
- **Handoff criteria:** Data quality scorecard green, duplicate resolution complete, segmentation-ready dataset
- **Stage owner accountability:** Data Engineering delivered a clean dataset with 2,134 unified profiles to Analytics

Day 7-10: Segment Stage (Analytics Team)

- **Deliverable:** Dynamic customer segments with behavioral scoring and propensity models

- **Execution:** Three primary cohorts emerged: "Flavor Adventurers" (34%), "Health-Conscious" (28%), "Price-Sensitive" (38%)
- **Quality metrics:** 89% prediction accuracy on purchase intent, segment stability >85% over 14 days
- **Handoff criteria:** Segment definitions documented, predictive models validated, business rules established
- **Stage owner accountability:** Analytics provided a segment charter with 2,847 customers categorized and scored

Day 11-13: Interpret Stage (Insights Lead)

- **Deliverable:** Insight brief with root-cause analysis and predictive recommendations
- **Execution:** "Flavor Adventurers" showed 3.4x higher lifetime value, but expressed concern about artificial ingredients in open-text feedback
- **Quality metrics:** Statistical significance >95% on key findings, qualitative analysis on 312 verbatim responses
- **Handoff criteria:** Insight brief approved, business implications quantified, recommendation priorities ranked
- **Stage owner accountability:** Insights Lead delivered actionable recommendations with $180K revenue opportunity identified

Day 14-16: Guide Stage (Functional Leaders)

- **Deliverable:** Decision log with approved actions, resource allocation, and success criteria

- **Execution:** Product team prioritized natural ingredient messaging; Marketing redirected $45K to educational content emphasizing organic berry sourcing
- **Quality metrics:** Decision latency 18 hours from insight brief to approved action plan
- **Handoff criteria:** Budget approved, resources assigned, activation timeline confirmed, success metrics agreed
- **Stage owner accountability:** CMO and Product Director signed off on $45K budget reallocation and messaging pivot

Day 17-19: Harness Stage (Marketing Tech + CRM Admin)

- **Deliverable:** Automated activation plan with personalized triggers and content delivery
- **Execution:** "Flavor Adventurer" segment received ingredient transparency email within 2 hours of purchase or website visit
- **Quality metrics:** 89% automation success rate, 47% email open rate (vs. 23% baseline), 156% CTR improvement
- **Handoff criteria:** Automation workflows live, personalization rules active, delivery confirmation systems operational
- **Stage owner accountability:** Marketing Tech confirmed 1,923 customers received personalized messaging

Day 20-21: Track Stage (Data Analytics + Finance)

- **Deliverable:** Performance dashboard with incremental lift measurement and ROI calculation

- **Execution:** Holdout group analysis showed the personalized segment achieved 3.7x higher conversion vs. control
- **Quality metrics:** $247K incremental revenue attributed to INSIGHT-driven personalization, 31% CAC reduction
- **Handoff criteria:** ROI verified, attribution methodology documented, learnings captured for next cycle
- **Stage owner accountability:** Finance validated $247K incremental revenue; Analytics documented insights for "Tropical Blend" launch

Framework Performance Metrics:

- **Learning velocity:** Insight-to-activation reduced from 14 days (previous campaigns) to 72 hours
- **Execution velocity:** Decision approval to live automation decreased from 5 days to 18 hours
- **Stage handoff reliability:** 100% on-time handoffs, zero escalations required
- **Business impact:** 4.2x ROI on framework implementation vs. traditional campaign management

Cross-Functional Acceleration: The systematic handoffs eliminated previous bottlenecks that stalled insights between departments. Marketing Ops, Data Engineering, and Analytics maintained 6-hour SLAs for stage transitions, while Functional Leaders reduced decision latency by pre-approving budget ranges for common optimization scenarios.

Compound Learning Effect: Insights captured during Berry Fusion informed subsequent product launches with 91% accuracy in predicting segment preferences. The systematic documentation enabled Northbridge to replicate successful tactics while avoiding previous mistakes, creating measurable acceleration with each INSIGHT Framework cycle.

This walkthrough demonstrated that disciplined stage execution with clear ownership transforms first-party data from scattered signals into systematic competitive advantage through measurable business outcomes.

Partner-Powered Momentum

Later that same week, Evelyn met with the retail partnerships team to address a long-standing challenge: how to extend the reach of Northbridge's data collection without relying solely on its own channels. The answer was to enlist trusted partners—distributors, retail chains, promotional companies, and event organizers—who already had high-value access to Northbridge's ideal customers.

By embedding Northbridge's capture points into partner-led marketing, the company can dramatically increase both the volume and diversity of first-party data it collects. Retail partners can host in-store demos with QR codes that link to a quick poll (see Chapter 7, "Consumer Insight Collection). After the visit, the system automatically sends follow-up communications—such as recipes, how-to guides, VIP experiences, or coupons—through the consumer's preferred channel (SMS, email, or social media). Event sponsors and channel partners can also run receipt-upload contests tied to co-branded offers.

Critically, you are not only gathering first-party data and building the customer relationship, but also attributing each signal to the correct partner channel. This lets Northbridge pinpoint which partnerships deliver the strongest ROI and strengthens future negotiations.

Sidebar: Partner-Powered Momentum in Practice

Promolytics provides you with the tools to enable partners to collect marketing data on your behalf, merging that partner-sourced first-party data into unified consumer profiles while preserving source attribution. This enables brands to measure channel-level ROI without compromising compliance. Promolytics can gather information via:

- **QR codes** on retail displays, event signage, packaging, print materials, TV advertising, or menus. See Appendix E: QR Code Statistics & Placement Playbook for the full set of placement ideas.
- **Trackable links** in partner emails, SMS campaigns, digital ads, or social posts
- **Partner-hosted microsites** with opt-in forms or embedded surveys
- **Receipt upload links** for rebate or loyalty redemption
- **First-party cookies** for consented web tracking
- **Poll and quiz links** inside email newsletters or product pages
- **Event registration links** for in-person or virtual experiences

The platform automates every stage of the process, from QR code capture and SMS or email click tracking to survey responses, receipt validation, and real-time activation triggers. Insights uncovered at one partner event can trigger a tailored offer across another channel within hours, keeping the flywheel spinning faster and smarter.

By the time Evelyn wrapped up the meeting, the team had outlined a six-week test with two regional retail partners and one national distributor. For the first time, the INSIGHT Framework would extend beyond Northbridge's direct touchpoints, accelerating the flywheel with external force.

Privacy Compliance: Partner data integration requires clear consent management across all INSIGHT stages. Ensure automated handoffs maintain consent status and source attribution. Sample protocol: 'All partner-collected data maintains original consent terms through every framework stage.' See Chapter 11 for complete compliance procedures.

Ritualizing the Framework: Turning Stages into Rhythm

Evelyn recognized that frameworks without a consistent execution cadence become theoretical exercises. To prevent the INSIGHT stages from becoming sporadic activities, she established a systematic review rhythm that locked each stage into repeatable timetables.

While Marketing Ops updated playbooks for each stage, Malik in Finance worked with Data Engineering to establish Service Level Agreement (SLA) targets for moving from Interpret to Harness within defined timeframes. The team learned that deadlines, not diagrams, kept the framework honest.

The cadence matrix below institutionalized this discipline, establishing review cycles from immediate post-campaign huddles to comprehensive annual assessments. When any scheduled review was missed, the entire insights loop lost momentum, and competitive advantage eroded.

The structure they developed operated on three distinct temporal layers, each serving a specific analytical purpose. Individual campaign reviews provided rapid tactical adjustments and immediate learning capture, while quarterly portfolio assessments enabled strategic pattern recognition across multiple initiatives. Annual reviews created the longest feedback loop, establishing year-over-year benchmarks and informing fundamental strategic pivots. This multi-tiered approach ensured that insights flowed continuously through the organization at the appropriate velocity and depth for each decision level.

Cadence Tier	Timing	INSIGHT Flow	Purpose & Key Outputs
1. Individual Campaign Review	First work week after the campaign has finished	1. Identify & normalize 2. Segment 3. Interpret 4. Guide decisions 5. Harness automations 6. Track	• Verify data capture is clean and complete. • Update audience segments with signals. • Surface one clear insight/storyline. • Approve optimizations and queue automations. • Circulate a quick-hit report; note any learning velocity issues.
2. Portfolio Review – All Active Campaigns (Quarterly)	One week after each fiscal quarter closes	Same six-step INSIGHT loop applied across the entire active-campaign set	• Spot cross-campaign anomalies and migration trends. • Publish a consolidated "Quarter in Review" deck. • Re-score segment health and growth. • Prioritize mid-quarter tests/budget shifts. • Refresh automation rules that span multiple campaigns.
3. Year-End & YOY Review – All Active Campaigns	Early January (or immediately after the fiscal year close)	INSIGHT loop & Year-over-Year comparison for recurring campaigns	• Compare key KPIs YOY (lift, ROI, redemption, LTV). • Highlight evergreen vs seasonal performance patterns. • Lock in strategic changes for the next annual planning cycle. • Archive data snapshots and audit compliance. • Issue an executive summary with a recommended 12-month roadmap.

INSIGHT Timeline

INSIGHT Framework Implementation Steps

Each INSIGHT stage required specific operational improvements to achieve the systematic handoffs and learning velocity Northbridge targeted. The following steps transformed the framework from concept to daily practice:

Identify Stage Enhancement:

- **Marketing Ops:** Standardized QR code placement protocols across 12 touchpoint types with real-time data ingestion, reducing capture lag from 4 hours to 15 minutes
- **Data Engineering:** Implemented automated identity key assignment,

eliminating manual profile creation delays

Normalize Stage Optimization:

- **Data Engineering:** Configured real-time ETL processing with 94% accuracy validation rules and duplicate resolution within 6 minutes of data receipt
- **Analytics**: Established data quality scorecards, preventing poor-quality data from advancing to segmentation

Segment-to-Interpret Acceleration:

- **Analytics Team**: Built automated segment health monitoring, detecting composition changes requiring re-analysis within 24 hours
- **Insights Lead**: Created insight brief templates, reducing interpretation time from 3 days to 8 hours

Guide-to-Harness Execution:

- **Marketing Tech**: Deployed trigger-based automation enabling approved decisions to activate customer touchpoints within 2 hours
- **Finance**: Established pre-approved budget ranges for common optimization scenarios, eliminating approval delays

Track-to-Identify Learning Loop:

- **Data Analytics**: Built automated performance dashboards, feeding results back into capture point optimization

- **Cross-functional team**: Implemented weekly INSIGHT reviews, ensuring learnings informed next cycle planning

These operational improvements reduced average cycle time from 21 days to 72 hours while maintaining quality standards and compliance requirements."

Executive Reflection:

Audit your organization's current data-to-decision process against the seven INSIGHT stages and identify which stage currently lacks clear ownership or defined handoff protocols. Where does your biggest bottleneck occur between capturing customer signals and taking action on insights? Which stage would benefit most from establishing formal Service Level Agreement (SLA) targets and measurement criteria within the next 30 days?

Chapter 9 Checklist: Operationalizing INSIGHT

- Document owners and KPIs for all seven stages.
- Add data quality and learning metrics to the dashboard.
- Pilot one automation that moves from Interpret to Harness within forty-eight hours.
- Schedule a quarterly audit to ensure each stage handoff—the formal transfer of deliverables and accountability between INSIGHT Framework stages with defined timing and quality criteria—occurs on time.
- Store all Track results in a searchable repository for future reviews.

The INSIGHT Framework provided Northbridge with systematic operations that accelerated learning velocity while maintaining quality and compliance. But as data volumes grew and customer expectations increased, manual interpretation and activation became bottlenecks that constrained the system's potential.

The next frontier was augmenting human intelligence with artificial intelligence—using machine learning to detect patterns humans might miss, predict customer behavior with greater accuracy, and automate real-time responses that kept pace with customer needs.

Chapter Ten

AI in the Loop:

From Human Insight to Machine Precision

AI implementation transforms from experimental analytics projects into systematic decision engines through real-time data pipelines, intelligent automation, and predictive customer engagement. This chapter presents a dual-track framework that leverages first-party signals to achieve an AI-powered competitive advantage through coordinated strategy and effective engineering execution.

We'll examine three high-impact AI use cases, including dynamic campaign optimization, real-time personalization, and generative creative content. We'll introduce infrastructure models that enable millisecond decision-making across channels and provide MLOps frameworks that ensure consistent AI velocity—the speed at which machine learning insights are deployed into live customer experiences. You'll learn to establish governance protocols that track model accuracy—the degree to which AI predictions translate into improved business outcomes—and ethical compliance across the complete AI lifecycle.

By the end of this chapter, you'll have operational blueprints for implementing AI-powered decision systems, feature engineering frameworks that maintain model performance while scaling data complexity, and measurement systems that prove AI ROI while accelerating continuous learning and automated optimization.

Northbridge: From Reactive to Predictive

Later in the week, Evelyn returns to the analytics lab, this time watching a cascade of data feeds from the latest campaign pour into the dashboard. What catches her attention isn't just the volume, it's the variety. QR code scans from in-store displays, receipt uploads from loyalty programs, link clicks from email newsletters, and even cookie-based interactions from websites are all being tracked in real-time. The mix reflects the breadth of channels Northbridge now taps into, each one carrying its own first-party data stream.

What had once taken days to reconcile now happened automatically, stitched together by algorithms trained to recognize the same consumer across multiple campaigns and channels. The system didn't just compile data; it interpreted it in real time.

- A spike in scans from the Midwest? The AI flagged it.
- A sudden shift in flavor preference among loyalty members? The model re-ranked segment priorities.
- A drop in conversion in a high-performing metro? The anomaly detector issued an alert to the sales team before store managers even became aware of it.

This wasn't just faster reporting; it was a shift from reactive analysis to **predictive action**. The AI could forecast inventory strain before a shortage occurred, identify which consumers were most likely to redeem a limited-time offer, and suggest creative variations most likely to improve response rates for each audience segment.

Real-World Application: Promolytics AI in Action

Today, Promolytics analyzes the data captured from each user experience, including QR scans, receipt uploads, survey responses, and link clicks, to identify performance patterns and provide suggestions for optimizing current campaigns or reorganizing future campaigns for stronger results. This means marketing teams can see what's working, spot underperforming elements, and adjust strategy before the next activation goes live.

Looking ahead, future AI and machine learning enhancements could allow Promolytics to move from *insightful analysis* to *predictive action*. Models can forecast consumer response by segment, detect early signs of campaign fatigue, recommend creative or offer changes in real-time, and even auto-trigger targeted activations within minutes of a consumer's engagement—keeping brands in step with shifting demand while the moment of intent is still fresh.

For Northbridge, the real opportunity isn't just gathering these signals—it's learning fast enough to act while the opportunity is still alive.

Why AI/ML Needs First-Party Data

First-party data, the information you collect directly from your customers' interactions, is the lifeblood of modern AI strategy. Why? Because AI and machine learning thrive on **relevance, immediacy, and trustworthiness** —hallmarks of first-party data. Unlike generic market research or third-party lists, first-party signals are *your* customers telling you what they intend to buy and what they actually purchase. It's the difference between navigating with a live GPS versus a paper map drawn a year ago. In practical terms, this means your models learn from *precisely* the right patterns. For example, Northbridge discovered that by mining its own purchase logs and site clickstreams, it could uncover product pairings and content that engaged its highest-value customers, insights no syndicated industry report could provide.

Equally important, first-party data creates a **holistic view** across touchpoints. When you unify data from your website, stores, social media, apps, and support lines, along with marketing insights directly from the customer, you gain a comprehensive view of the entire customer journey, rather than fragmented pieces. An AI can then answer those golden "why" questions: *Why are customers in Re-*

gion A churning more? Which messages lead not just to clicks but repeat purchases? Patterns emerge only when formerly siloed data comes together, turning your first-party dataset into an organizational "eye" that sees the whole picture. This integration becomes a **competitive moat**—a unique asset that competitors can't easily copy. Your customer behaviors and preferences are proprietary; models trained on them can predict and personalize in ways a rival using generic data simply cannot.

In the age of privacy, this advantage is even more substantial. First-party data is collected with customer consent, making it privacy-compliant by design. Brands that lean into this approach build trust and sidestep the risks associated with third-party data, which is increasingly scarce in an evolving data privacy landscape. In short, **AI requires first-party data because it's richer, more timely, and more ethically sound**. It's no coincidence that companies winning with AI (from streaming services to retail leaders) are masters at harnessing their data, turning consent-based customer information into smarter predictions and personalized experiences. As Northbridge's leadership came to realize, first-party data isn't just another input for AI; it's a strategic asset that grows more valuable with use, forming the foundation of an AI-driven competitive edge.

The Dual Track Framework

To turn live signals into instant advantage, Northbridge adopted a dual-track approach: one **strategy track** for the "why and what," and one **engineering track** for the "how." This ensures that high-level business goals align with technical execution. In the strategy track, we'll explore three high-impact AI use cases that first-party data enables. In the engineering track, we'll outline how to build the data and ML pipeline to support those use cases. Both tracks run in parallel; executives guide the vision and use cases, while technical teams build the capability.

Strategy Track: Three High-Impact AI Use Cases

1. Dynamic Campaign Optimization

Why it matters: Marketing dollars evaporate when the channel, audience, or creative is frozen for weeks. A dynamic campaign engine reallocates spend every few minutes, pushing budget toward the ads, audiences, and geographies that are currently proving incremental lift. Think of it as an autopilot that shifts your ad fleet to the clearest skies and most profitable destinations, before the competition's manual report even loads.

How it works: Models ingest first-party signals (real-time coupon redemptions, receipt-validated sales, fatigue scores) and predict marginal return for every live campaign node. If the expected coupon redemption rate drops below a threshold, bids decrease; if a microsegment surges, the budget and impressions increase accordingly. For Northbridge, the engine cut underperforming Instagram placements within two hours. It redeployed the freed budget to regional TikTok ads that were converting at a 3× lift, optimizing $ 18,000 in spend during the first week.

2. Real-Time Content Personalization

Why it matters: Static homepages and one-size-fits-all emails treat loyalists and first-timers the same, eroding engagement. Real-time personalization swaps modules, headlines, or entire offers based on an individual's *immediate* behavior. For a storefront analogy: it's like a clerk changing the front-window display the moment a VIP walks by, except it happens for millions of shoppers simultaneously.

How it works: A lightweight inference service scores every visitor in milliseconds using first-party features such as live browse path, recent purchases, and stated preferences. The CMS then assembles a page from a library of content blocks:

a recipe article for the health-focused cohort and a bundle discount banner for value seekers. Northbridge experienced a decrease in bounce rate on mobile after implementing this component-level personalization, demonstrating that displaying the *right* message, not just any message, yields immediate results.

3. Generative Creative Content:

Why it matters: In marketing, fresh and personalized content drives engagement, but creating it manually is slow and expensive. AI can act like an always-on creative assistant, generating copy, images, or designs tailored to your customers. Think of it as a digital brainstorming team that creates individualized ads or emails for each customer segment in minutes, or a billboard with endless variants, each tailored to the person standing in front of it.

How it works: AI models (particularly generative models like GPT or image generators) learn from your first-party data about customer preferences and past campaign performance. They then produce new creative variations and even optimize them on the fly. For example, Northbridge's marketing team might use a generative AI to create dozens of email subject lines personalized to different customer segments (new customers get a "Welcome" tone, loyal customers see a "Thank You" message with their first name, inactive customers get a "We Miss You" note). Similarly, image generation models could produce dynamic ad banners showing the specific product a customer browsed but didn't buy. The technical team could feed the AI a prompt like "Produce an Instagram ad image highlighting Product X, with a style appealing to our 20–30 age segment (based on our first-party engagement data)." The AI then outputs a set of on-brand visuals. Over time, the system learns which creative variants perform best for which audiences by checking first-party performance metrics (click-through rates, conversions) and can refine content accordingly. The net effect is one-to-one marketing at scale: *every* customer sees messaging tuned to them, created in real-time by AI. This use case transforms first-party data (such as user preferences and past behaviors) into bespoke creatives that would be impossible to produce manually at the same speed or volume.

From a technical standpoint, Northbridge had two options for implementing generative AI. For rapid deployment, they could leverage pre-trained models like GPT or DALL-E via API, accepting standard outputs but maintaining simpler infrastructure. For deeper customization, they could fine-tune open-source models using their first-party data, historical high-performing creative assets, customer engagement patterns, and brand guidelines, thereby creating a model that inherently understood Northbridge's unique voice and visual style.

The team chose a hybrid approach: API-based generation for initial creative variants, followed by fine-tuning based on performance data collected through their first-party tracking. This allowed them to maintain brand consistency while learning which generated elements resonated with specific segments. Critical to this process were brand safety controls, automated reviews checking for inappropriate content, trademark conflicts, and message alignment before any AI-generated creative went live.

These strategy-track use cases illustrate the *why*: dynamic campaign optimization, real-time personalization, and generative creative all feed the bottom line. But none of these would work without the *how*: a robust engineering foundation that can capture signals and feed them to AI models quickly and reliably. Next, we switch to the engineering track, the technical blueprint that makes the above scenarios possible.

Executive Summary - Engineering Foundation: Building AI that acts on customer signals requires three core systems: real-time data pipelines that capture and process events within milliseconds, feature stores that convert raw data into AI-ready inputs, and MLOps processes that keep models accurate over time. Without this foundation, AI becomes a science experiment rather than a business driver. Investment timeline: 3-6 months for basic capability, 12-18 months for enterprise scale.

Engineering Track: Data to Deployment (Pipeline, Features, and MLOps)

Building AI with first-party data requires a robust real-time pipeline, a well-designed feature store, and disciplined MLOps. Northbridge's engineering focus covers **real-time data pipelines**, **feature store design**, and **model operations cadence**. This path enables data to be transformed from raw events to live predictions.[1] Building this infrastructure required careful planning. Northbridge's architecture team established that their initial deployment would need to support 10 million events daily, scaling to 100 million during peak seasons. They also implemented comprehensive fallback mechanisms: cached predictions when models were unavailable, rule-based decisions when latency spiked, and graceful degradation that prioritized system stability over perfect personalization. As their CTO noted, "Better to show a good-enough experience to everyone than a perfect experience that crashes under load."

Real-Time Data Pipeline

First-party signals create value only when they are captured and processed quickly. Every touchpoint—web, app, in-store, tastings, events, print, TV—should send events into a streaming backbone like Kafka. Utilize tools like Flink to aggregate data in real-time. The pipeline should handle both streaming and batch feeds. Set clear latency targets: sub-second speed for real-time optimization, and a few seconds where speed is less critical. Specifically, Northbridge targeted:

---

1. MLOps/ModelOps practices: Google MLOps (last reviewed Aug 28, 2024); TFX at Google (KDD 2017); AWS Model Monitor/Clarify (drift, bias). https://cloud.google.com/architecture/mlops-continuous-delivery-and-automation-pipelines-in-machine-learning

- 100ms for critical path decisions (bid adjustments, content swaps)
- 500ms for personalization calls
- 2-3 seconds for complex scoring operations

The system was designed to handle 50,000 events per second during peak campaigns, with automatic scaling to 200,000 events per second during major promotions. Equally important were the fallback strategies: when real-time scoring failed, the system defaulted to the last known good score; when streaming pipelines lagged, batch predictions from the previous hour took over; and when personalization engines couldn't respond fast enough, pre-computed segment-level defaults ensured customers always saw relevant content rather than errors.

These expectations ensure optimization engines can act before budget is wasted. Core requirements include:

- **Inline data hygiene**: filter invalid traffic, validate schema/timestamps, and keep clocks in sync to ensure the stream is clean upon arrival.
- **Identity & keys**: Each event—impression, click, purchase, survey, QR, link, or support chat—should be tagged with a durable customer identifier and contextual IDs, such as SKU, campaign, or content. Update the customer's unique profile with these events. This stitching keeps behavior coherent across apps, sites, and stores.
- **Streaming aggregation:** compute rolling metrics in windows—such as 5-minute intervals—so models and alerts use the latest data. Trigger actions when thresholds are crossed, for example, when a conversion rate drops suddenly. The enriched event stream updates the feature store and calls live model endpoints. Automated quality gates, like spike alerts and anomaly detection, protect against bad data. Ultimately, the system anticipates decisions ahead of market changes.

Executive Summary - Real-Time Pipeline: Think of this as your AI nervous system. Every customer touchpoint sends signals that flow through a central processing backbone, creating a live view of customer behavior. The technical complexity is significant, but the business impact is immediate: campaigns that optimize themselves, inventory that adjusts to demand, and customer experiences that improve in real-time. ROI typically appears within 30 to 60 days of deployment.

Feature Store Design

After event streams flow in, the next challenge is to make them usable for ML. Convert raw data into features, which are either numerical or categorical data that models learn from. Examples: average order value or logins in the past 7, 30, or 90 days, and preferred category. A feature store prevents teams from re-inventing features. It supports two layers:

- **Offline store** (data lake/warehouse) for historical training and batch inference.

- **Online store** (low-latency key-value) the latest feature values for real-time inference. For instance, compute a nightly churn score from 90-day behavior and store it offline. When a new event happens (like a support ticket), update 'recent_support_tickets' online so the model can re-score immediately. Requirements include using the same definitions both offline and online, allowing for low-latency reads, and tracking feature and model versions that drive decisions.

In practice, a mix of SQL and Python jobs—both streaming and scheduled—create features. For example, a nightly job updates 30-day spend and last-purchase date. The job then syncs to the online store to serve these features.

Executive Summary - Feature Store: This is your AI memory bank. It stores both historical patterns (for training models) and live customer data (for instant decisions). Without it, each AI project rebuilds the same data foundations, wasting months of development time. A well-designed feature store accelerates the development of new AI use cases from months to weeks and ensures consistent definitions across all models.

Northbridge set explicit performance benchmarks for its feature store: online feature retrieval in under 10 milliseconds for 99% of requests, batch feature computation completing within 2 hours for daily features, and streaming features updating within 30 seconds of event occurrence. These targets ensured that even during Black Friday traffic spikes, when event volumes could reach ten times normal levels, the AI systems maintained their responsiveness.

Executive Summary - Performance Standards: Speed matters in AI. Sub-second response times enable real-time personalization; anything slower becomes batch reporting. These performance targets aren't technical nice-to-haves—they determine whether your AI can influence customer behavior in the moment or only analyze it after the fact. Set clear latency requirements before building, not after.

MLOps Cadence

Models in production require ongoing attention, including training, deployment, monitoring, and continuous improvement. Create a regular schedule for continuous integration, delivery, and training. Watch for **data or feature drift** and check **performance**. Before rolling out a new model, test it on holdout data or with controlled experiments. Automate retraining triggers—based on time or drift—and use containers and infrastructure as code to manage serving. Line up model upgrades with the business calendar, for example, tuning before busy seasons. A sample schedule: **daily** data checks, **weekly** reviews, **monthly** retraining or on drift, and **quarterly** feature updates (frequency should be adjusted based on business criticality and data velocity). This consistency ensures that systems continue to deliver value after launch.

Executive Summary - MLOps: AI models are like high-performance cars—they require regular maintenance to stay effective. Customer behavior changes, market conditions shift, and models degrade without updates. MLOps creates the discipline to monitor, retrain, and deploy models systematically. Companies that skip this step often see AI performance drop 20-40% within six months of deployment.

Tooling Map: Aligning INSIGHT Stages to AI/ML Tools

Having discussed strategy and engineering, it helps to see a one-page view of how everything ties together. Northbridge utilizes the INSIGHT framework (Identify, Normalize, Segment, Interpret, Guide, Harness, Track) to transform raw data into actionable insights. Below is a tooling map that aligns each INSIGHT stage to representative AI/ML tools or techniques. This is where narrative meets nuts-and-bolts; each stage of the journey from signal to sale has enabling technology.

The following mapping illustrates how each INSIGHT stage corresponds to specific AI/ML capabilities. This isn't just theoretical; Northbridge utilized this exact framework to build its AI infrastructure, ensuring that every technology investment directly supported its first-party data strategy.

What emerges from this comprehensive tooling approach is a technology stack that operates as an integrated system rather than isolated point solutions. Each stage's tools are deliberately chosen to feed seamlessly into the next, creating data pipelines that maintain context and momentum throughout the entire customer intelligence cycle. The real power lies not in any single tool, but in how streaming data flows from capture systems through ML models to real-time decision engines, creating an unbroken chain from customer signal to automated action. This architectural coherence ensures that insights don't get trapped in departmental silos but instead flow dynamically across the organization to drive immediate business value.

INSIGHT Stage	AI/ML Tools & Techniques
Identify: Capture first-party signals at the source. *What customer signals can we capture?*	Data capture tools (web/app analytics SDKs, IoT sensors, and POS systems) are used to log events. Streaming ingestion frameworks (e.g., Kafka, cloud event hubs) to collect clicks, views, and transactions in real time. Tag managers and CDPs (Customer Data Platforms, Promolytics) to ensure all touchpoints are instrumented for data collection.
Normalize: Make data usable and consistent. *How do we unify and clean the data?*	ETL/ELT pipelines (batch jobs or real-time processors) for cleaning, deduplicating, and joining data streams. Data quality tools (automated validation scripts, anomaly detection on incoming data) to catch errors. Identity resolution systems merge disparate records into a single customer profile. Feature engineering pipelines (using SQL or Spark) to transform raw events into model-ready features.
Segment: Group customers by behavior or value. *Which groups exhibit shared patterns?*	Analytics & ML clustering techniques: for example, SQL-based segmentation in a data warehouse (e.g., query a loyalty database for high spenders), or unsupervised algorithms like K-means to discover natural clusters. Business rules in CDPs/CRMs for rule-based segments (e.g., "all customers who purchased >3 times in 90 days"). This stage can also utilize classification models (e.g., a propensity model that labels users as "likely to churn" versus "not likely").
Interpret: Derive insights and predict outcomes. *Why do segments behave this way, and what will they do next?*	Machine learning training libraries (TensorFlow, PyTorch, scikit-learn) to build predictive models on historical first-party data – e.g., churn predictors, lifetime value (LTV) estimators, demand forecasts. Notebooks and analytics tools (Jupyter, RStudio) for deep-dive analysis and visualization to explain segment behavior. Explainable AI techniques (SHAP values, LIME) to interpret model drivers – for example, identifying which features (age, usage frequency, price sensitivity) contribute most to a churn prediction or a pricing decision.

INSIGHT Framework AI/ML Tools Mapping (Part 1)

INSIGHT Stage	AI/ML Tools & Techniques
Guide: Turn Insights into Business Decisions. *Which business decisions must change?*	Decision-support and visualization tools to surface model findings in plain English for decision-makers. This could be BI dashboards highlighting, for example, changes in ROAS by channel or alerts about a drop in feature usage. What-if simulators (spreadsheets or lightweight apps) to test scenarios – e.g., "If we shift 10% of budget from Channel A to Channel B, what is the projected revenue impact?" Optimization solvers (from simple linear programming to reinforcement learning agents) can recommend the best allocation of budget, offers, or inventory based on model outputs. Collaboration platforms (shared documents, project boards, Slack/Teams integrations) ensure that insights are turned into agreed-upon actions across marketing, sales, product, and other departments.
Harness: Operationalize insights in real time. *How do we act on the insight immediately?*	Real-time inference and automation tools: model-serving infrastructure (open-source stacks like TensorFlow Serving, FastAPI microservices, or cloud serverless functions) to deliver ML predictions on demand – e.g., a churn probability score, a price recommendation, or a content personalization slot for User 123. Orchestration layers or rules engines then inject the AI's decision into the live environment: insert the selected **content, price, or bid adjustment** into the website, app, email, or ad platform within milliseconds. Marketing automation and notification systems listen for triggers (via API or webhook) – for instance, if a pricing model flags that a product's price should drop, the system applies the new price on the website and sends an alert to the pricing manager. Event-driven workflow tools (like Airflow or cloud event buses) glue everything together so that every prediction leads to an action without human delay.
Track: Measure impact and keep learning. *Did our action create value?*	Analytics, experimentation, and monitoring tools close the loop. A/B testing or feature-flag platforms hold out control groups to rigorously measure lift from AI-driven changes. KPI dashboards roll up outcomes, such as incremental revenue, engagement lift, and retention rates for the test versus the control groups, providing executives with an evidence-backed "scoreboard." In parallel, ML monitoring services (such as MLflow, Evidently, WhyLogs, or custom scripts) track model performance for drift, bias, or latency issues, raising alerts if something deviates from the expected track (for example, if the churn model's predictions cease to correlate with actual churn outcomes). Crucially, all results feed back into the data lake tagged with the actions taken (which offer was shown, which price was set, etc.), so the next **Identify→Segment** cycle has even richer first-party truth to learn from.

INSIGHT Framework AI/ML Tools Mapping (Part 2)

Note: The tools and techniques in the chart are examples. Northbridge maintained a vendor-neutral and flexible approach, opting for open-source components and modular cloud services that best fit their needs. The key is that each stage of INSIGHT has an owner and a technology process to ensure data flows smoothly from raw signal to final business value.

Cross-Functional Implementation: AI/ML Accountability Matrix

AI initiatives fail when they become isolated technical projects rather than integrated business capabilities. Unlike traditional analytics, which reside within data teams, AI systems that act on customer signals in real-time require coordination across every function that touches the customer experience. When a churn prediction model flags a high-value customer, marketing must have pre-approved offers ready, customer success needs escalation protocols in place, and legal must have verified that all interventions comply with privacy regulations. The following accountability matrix ensures that AI capabilities become organizational muscle memory rather than one-off experiments.

Marketing Operations: Owns campaign optimization models, creative performance tracking, and A/B testing infrastructure. Responsible for setting business rules and performance thresholds that trigger automated actions.

Data Engineering: Manages real-time pipelines, feature store maintenance, and model serving infrastructure. Ensures sub-second response times and maintains fallback systems during peak loads.

Data Science: Develops and validates models, monitors drift and bias, and provides explainability reports. Conducts regular model audits and retraining schedules.

Legal/Compliance: Reviews all AI use cases for privacy compliance, bias risks, and regulatory alignment. Approves data retention policies and customer communication about AI usage.

Customer Success: Monitors AI-driven customer interactions for satisfaction impact and provides feedback on model predictions versus actual customer behavior.

Governance & Ethics

These governance practices aren't optional extras; they're competitive necessities. Companies that deploy AI without proper oversight often face public relations disasters, regulatory penalties, or erosion of customer trust that can take years to repair.

Executive Summary - AI governance isn't about slowing innovation—it's about preventing disasters that kill AI programs entirely. Public bias scandals, regulatory violations, and erosion of customer trust can shut down AI initiatives overnight. Smart governance establishes guardrails that enable faster and more confident AI deployment while protecting brand reputation and ensuring regulatory compliance.

Building AI on customer data is powerful, but with great power comes great responsibility. Northbridge instituted strong governance and ethics practices to ensure its AI/ML initiatives remain fair, transparent, and accountable. (Chapter 11 will explore the critical privacy and permission frameworks that underpin these systems.) Four pillars of this governance are bias monitoring, model drift management, explainability, and cross-functional oversight:

- **Bias Monitoring:** First-party data reflects your customers, but it can also reflect your company's historical biases. Northbridge's data science and compliance teams established routines to regularly review models for unintended bias. For example, if the content-personalization model consistently offers lower discounts to customers from a specific region or demographic, that's a red flag to investigate. Tools that slice model outcomes by group (e.g., by age, gender, location) are used to detect disparities. The team also scrutinizes training data during the **Normal-**

ize/Interpret stages – removing or down-weighting features that could proxy for protected attributes (such as ZIP code acting as a proxy for race, for example). The goal is to **ensure fairness** in outcomes so that the AI doesn't inadvertently discriminate or harm the brand's reputation. Any biased findings prompt a retraining of the model with adjustments (or even a decision *not* to automate a decision if it can't be made fair). This proactive stance aligns with a best practice: rigorously examining training data to prevent the embedding of real-world biases into AI algorithms.

- **Model Drift Management:** Even a well-trained model can go stale. Customer behavior today may not be the same as it was six months ago (new competitors, seasonality, regulatory changes – you name it). Northbridge combats this through continuous **model monitoring**. They track live performance metrics – for instance, the churn model's precision and recall are logged each week. If these metrics start degrading beyond a set threshold, or if the statistical properties of incoming data shift (for example, average session duration increases due to a site redesign), automated alerts are triggered. Think of model drift like milk nearing its expiration – you check the date and sniff periodically. When something seems "off," the data science team intervenes: retraining the model with fresh data, or even rolling back to a previous model version if a new one misbehaves. To do this smoothly, Northbridge invested in **MLOps pipelines** that can retrain and redeploy models with minimal manual work. Regular retraining schedules are also in place (e.g., every month or quarter, depending on the rate of data change) to ensure models stay current. In short, governance here means *never assuming a model stays good forever* – it's about instituting a lifecycle where models are born, regularly evaluated, and sometimes gracefully retired or reborn to match the changing world.

- **Explainability and Transparency:** Gaining buy-in for AI initiatives at the executive level (and with customers) requires trust. Northbridge addresses this by making AI decisions as explainable as possible. For any given prediction – say, the churn model flags Customer Jane Doe at 80% risk – the system also produces an explanation such as: *"Key factors: No logins in 30 days, three recent support complaints, downgrade from premium tier."* These explanations come from techniques like SHAP values (which can rank feature importance for that prediction) or simple rules for specific models. The technical team builds **explainability into the interface** that business teams use: a marketer sees not just *who* is high-risk, but *why* the model thinks so. This fosters understanding and enables more innovative interventions (if the reason is "support complaints," maybe a personal call is better than a generic discount email). Moreover, Northbridge is transparent about its use of AI with customers whenever appropriate. For example, if **product recommendations** are being personalized, they set guardrails to avoid sensitive inferences and consider disclosing, "Suggestions are based on your activity and the preferences you shared," to avoid customer backlash. Internally, **model documentation** (akin to "model cards") is maintained for each AI system, recording its intended use, limitations, and evaluation results on fairness and accuracy. This documentation is reviewed in governance meetings. The mantra here is that an AI model shouldn't be a black box to those relying on it; shine light inside the box so stakeholders are comfortable and informed.

- **Cross-Functional Review Cadence:** AI in business isn't just a tech project; it touches marketing, sales, legal, PR, and beyond. Northbridge established a cross-functional AI Governance Council that meets on a regular schedule (e.g., bi-monthly). In these meetings, data scientists present recent model performance and any incidents (like "we noticed the model making odd recommendations for Segment X last week"). Compliance officers review any new regulatory considerations (say, new privacy laws) to ensure data usage stays within legal bounds. Marketing and sales leaders provide feedback on whether the AI insights align with their observations in the field. This type of forum ensures accountability and fosters a **shared understanding of the issues at hand**. It's essentially a team sport: if the churn model is generating retention actions, the marketing VP, the customer success lead, and the data science head are all in the loop on its impact and any issues. The cross-functional council also decides on any significant changes – for example, approving the rollout of a new AI-driven initiative (such as using AI to automatically generate creative content for an upcoming campaign) only after considering reputational and customer experience aspects. By having a scheduled review (akin to an AI/ML steering committee), Northbridge prevents the siloed development of AI that might go rogue or fail to align with business strategy. It's a practice that echoes what leading companies do, with some even having formal AI ethics boards to govern the comprehensive use of AI. The takeaway: Governance isn't a one-time checklist; it's an ongoing culture of oversight and accountability. From bias scans to executive reviews, Northbridge treats its AI not as a magic box, but as a set of business processes that must be evaluated and improved just like any other – ensuring the "instant advantages" from AI never come at the expense of ethics, customer trust, or regulatory compliance.

Northbridge Case Study: 7-Step Live Data to Revenue Walkthrough

To cement how all these pieces come together, let's walk through a simplified end-to-end scenario. Imagine Northbridge offers a premium subscription service and wants to reduce customer churn – a clear revenue-saving goal. Here's the step-by-step journey from data capture to business lift, aligned with the INSIGHT framework:

1. **Identify:** Northbridge's website is instrumented to capture early signals

of declining engagement. For example, the moment a user clicks "Cancel Subscription," that event is logged, and an in-app exit survey asks why they're leaving. At the same time, ongoing usage data (e.g., last login date, feature usage frequency) and customer service data (e.g., open support tickets) are streaming in for every user. These first-party signals – *haven't logged in 30 days, clicked 'Cancel subscription', opened three support tickets this month* – are identified in real time and immediately sent into the data pipeline.

2. **Normalize:** As soon as those raw events stream in, Northbridge's data platform cleans and merges them into a unified customer profile. The user's identity is resolved across systems (for instance, the app's anonymous user ID is matched to the master CRM record for Jane Doe). The cancellation-click event, survey response text, and latest usage metrics are all aggregated into a consistent format. Think of this step as updating a customer's status in the central database: *Jane Doe – Premium subscriber; last login 45 days ago; 3 support tickets in the previous month; clicked "Cancel" on 2025-07-15; survey reason: "Not enough new content."* Every data point is timestamped and standardized (time zones reconciled, text fields cleaned, etc.) so that they're ready for analysis. This normalization occurs via an automated ETL job or streaming process that updates the latest attributes in Jane's feature store profile almost instantly.

3. **Segment:** Almost immediately, Jane Doe falls into a pre-defined high-risk segment. Northbridge's analytics team has set up dynamic segments (some rule-based, some model-driven) to categorize users based on their behavior. In this case, Jane meets the criteria for **"Long-time premium subscribers with a recent drop in activity and signs of cancellation intent."** The system tags her as part of the **"At-Risk Premium"** cohort. Part of this is powered by a predictive model as well – Jane's data triggers a churn propensity model, which scores her with a 0.85 probability of churn (above the 0.80 threshold for high risk). That

score is another first-party signal that lands her in the high-risk bucket. Additionally, analytics reveal that there are over 200 other customers *just like her* this week. These segments are living cohorts that update as data flows in – if Jane logs back in tomorrow and watches five videos, her risk score might drop, and she could exit the segment; but right now, she and the others are flagged as needing attention.

4. **Interpret:** Now the team asks **why** these premium users are on the brink of churning. Here's where data analysis and AI join forces to find insights. They notice a pattern: many of the at-risk users (like Jane) answered the exit survey with comments like "content not relevant" or "too expensive for what I get." An NLP model quickly parses the free-text survey answers and highlights common themes – "no new features" and "cost increase" top the list. Meanwhile, the churn prediction model's explainability output shows the top factors driving Jane's churn risk are **"low feature usage in the last 30 days"** and **"recent price increase on plan."** In essence, the insight emerges: these valuable customers haven't seen new content or features to justify a recent price hike, so they're losing faith in the service's value. The data team also projects what might happen next if we do nothing – approximately 60% of this high-risk cohort is expected to *cancel within the next month*. This Interpret stage combines human and machine intelligence to form a narrative hypothesis: *Premium users are disengaging due to a perceived lack of new value after a price increase, and many are about to leave if we don't act.*

5. **Guide:** Armed with that insight, Northbridge's cross-functional team convenes (virtually, of course) and formulates a game plan to change the outcome. They decide on two immediate actions to re-demonstrate value to these at-risk customers: **(a)** offer a one-time loyalty discount or account credit to acknowledge the price increase pain, and **(b)** fast-track the release of a new premium feature (or some exclusive content drop) that this cohort has been clamoring for. In the weekly strategy meet-

ing, the product manager commits to pulling a planned feature from next quarter into this month's release, and the marketing lead allocates budget for the loyalty discounts. Essentially, the business is saying: *"Let's show these customers love now so we keep them."* This Guide stage is where the insight turns into a concrete action plan – deciding *who* to target (the "At-Risk Premium" segment), *what* to offer them (a discount + new feature), *how* to reach them (email, in-app message, and perhaps personal phone calls for the highest-value ones), and *when* (immediately, before they churn). Everything is documented as a mini-campaign and playbook, ensuring clarity for all teams involved.

6. **Harness:** Now execution kicks in – primarily via automation, because the goal is to act in real time. Northbridge's CRM and marketing automation systems are integrated with the churn model's outputs and the segment definitions. The moment Jane Smith was identified as high-risk, behind the scenes, an automated workflow had already queued up a personalized email and in-app message for her. As soon as the team gives the green light on the offer, the campaign triggers: Jane receives an email *that very day* saying, *"We noticed you haven't been enjoying Northbridge Premium lately. Here's a 20% loyalty discount on next month's fee, and a sneak peek of new content we're launching just for members like you."* When Jane opens the app, she is greeted by a pop-up showcasing a brand-new premium content section that was added this week. These actions were implemented through if-then rules in the marketing platform (e.g., if churn_score > 0.8 and plan_status = 'Premium', then send offer X and display Feature Y). For the roughly 200 others in the segment, the system does the same automatically – perhaps staggering the messages slightly to avoid overwhelming customer support if many people respond. Importantly, Northbridge is careful to include a **holdout group**: a small, randomly selected portion of the at-risk cohort (say 10%) does **not** receive the offer immediately. They'll serve as a comparison to measure the offer's effectiveness truly. This Harness stage is where AI

meets the customer in real-time: predictions and plans are deployed as personalized actions (emails, app content, account flags for the support team) within hours of the initial churn signal. It's live first-party data being turned directly into live intervention.

7. **Track:** Over the next several weeks, Northbridge will diligently track the performance of this savings campaign. Dashboards update daily: **How many at-risk users used the discount and stayed? Are engagement metrics improving for them?** The team compares the outcomes of those who received the offer with those of the holdout group who didn't. The results speak loud and clear. Historically, 60% of this high-risk cohort would have churned within a month; after the campaign, only 30% ended up churning – Northbridge effectively halved the churn rate for that group. The difference between the holdout and the contacted group represents *incremental lift* attributable to the AI-driven intervention. In our example, that might translate to roughly **$200,000 in revenue retained** that would have been lost. Secondary metrics are also moving in the right direction: premium user satisfaction scores are ticking upward, and support ticket volume, which includes complaints about "value for price," is decreasing. All these outcomes are fed back into the next cycle. The data on who stayed vs. who churned is used to **retrain the churn model**, making it more accurate for next time (perhaps the model learns that customers who responded to the discount are less likely to churn in the future). The success of the new feature release validates the product team's decision and encourages more investment in content. Finally, the team holds a brief post-mortem (as part of the cross-functional council review) to document what worked and update the playbook for future retention efforts.

Executive Summary - Live Data to Revenue: This seven-step process demonstrates how first-party data can be leveraged to generate revenue protection in real-time. The key insight: AI doesn't replace human judgment—it accelerates it. While the technology is complex, the business logic is straightforward: capture signals, understand patterns, act promptly, measure results, and continually improve.

By walking through these seven steps, we see how Northbridge "spins the flywheel"; from capturing raw data to taking action to measuring impact, all within a tight loop. Every step is both enabled by first-party data and enriched by AI/ML tools, demonstrating the power of turning live signals into immediate business advantage.

Executive Reflection:

Looking at Northbridge's AI implementation - real-time personalization, churn prediction, and generative creative content - which of these capabilities would move your campaign results the furthest?

Chapter 10 Checklist: AI/ML in the Flywheel

- Map each INSIGHT stage to AI/ML capabilities and note any data dependencies.
- Select one high-impact use case (campaign optimization, personalization, or generative content) for pilot implementation.
- Establish performance benchmarks: latency targets, accuracy thresholds, and business impact metrics.
- Test one AI-powered prediction against 6 months of historical campaign data to validate accuracy.
- Define cross-functional ownership for each AI capability to ensure accountability.
- Set up governance review schedule with legal, marketing, and data

science stakeholders.

- Document all AI/ML data inputs, outputs, and retention timelines for Chapter 11 privacy review.

Real-time intelligence raised the bar for accountability. Every AI decision carried legal weight, from the creative content generated to the customer segments targeted to the personal data processed. Evelyn closed the session with a directive that would reshape how Northbridge approached customer data: "If we're going to act on signals in milliseconds, we need legal frameworks that protect customers in microseconds."

As the session ended, Jasmine Lee, Northbridge's Chief Legal Officer, prepared to step forward. The privacy landscape was shifting faster than the technology, and winning companies would be those that turned compliance from a constraint into a competitive advantage. The next chapter reveals that strategic privacy frameworks not only protect companies but also accelerate growth by building customer trust, which fuels first-party data collection.

Chapter Eleven

The Consent Framework:

When Privacy by Design Powers Customer Trust

Privacy compliance transforms from reactive legal scrambling into proactive competitive advantage through systematic consent management, transparent data practices, and trust-building customer relationships. This chapter presents the privacy-by-design framework, which transforms regulatory requirements into business assets through coordinated legal, technical, and marketing execution.

We'll examine the evolving U.S. privacy landscape, including state-specific requirements, browser restrictions, and regulations governing automated decision-making. We will introduce governance models that streamline compliance across partners and channels, and provide consent frameworks that ensure consistent privacy velocity, the speed at which privacy decisions are implemented across all customer touchpoints. You'll learn to establish trust protocols that track consent quality, the degree to which customer permissions translate into engaged, loyal relationships, and compliance effectiveness across the complete data lifecycle.

By the end of this chapter, you'll have operational blueprints for implementing privacy-by-design processes, consent management frameworks that maintain legal compliance while enhancing customer trust, and measurement systems that prove privacy ROI while accelerating partnership approvals and customer engagement quality.

Northbridge: Privacy by Design

Tuesday morning. Jasmine Lee, Northbridge's legal counsel, walked into the executive meeting carrying a thick binder labeled *State Privacy Statutes*. She set it on the table with a solid thud.

"We need to talk about privacy and permission," she says. "A perfect storm is forming as states layer new rules with different definitions and obligations." Evelyn nods—this conversation was inevitable. The company's data program had grown fast, and while AI models delivered sharper insights every week, those insights were only as strong as the trust behind them.

> **Definition:** In this chapter, *privacy* means how Northbridge collects, uses, stores, and shares customer data; *permission* means the customer's explicit, revocable authorization for a specific use.

The Rising Tide of US Privacy Laws: Why regulators keep raising the bar

What began as a single-state concern is now a rapidly evolving patchwork. New California regulations on Automated Decision-Making Technology (ADMT), risk assessments, and cybersecurity audits were approved in July 2025 with compliance windows running into 2027–2028. States continue to layer opt-out requirements, including Global Privacy Control signals that must be honored in California, Colorado (effective July 1, 2024), Connecticut (effective January 1, 2025), and New Jersey (by July 15, 2025). Meanwhile, browsers are limiting tracking unevenly: Safari and Firefox block most third-party cookies by default, while Chrome has adopted a user-choice model and continues to roll out Privacy Sandbox APIs. The upshot: design for consent and UOOMs (Universal Opt-Out Mechanisms) now, and modernize measurement without waiting for a universal cutoff. Appendix G: US Privacy Timeline & 2025 State Matrix illustrates the rapid evolution of the privacy landscape, highlighting why Northbridge can't afford a wait-and-see approach. As you scan each year, note the pattern: every statute tightens consent rules, narrows data use, or expands consumer rights. Taken together, they form the storm Jasmine warned about.

Browsers are continually tightening their tracking, but not uniformly. Safari (ITP) and Firefox (ETP / Total Cookie Protection) block most third-party cookies by default. Chrome shifted in 2024 from a planned third-party cookie deprecation to a user-choice model, continuing to develop Privacy Sandbox APIs while maintaining third-party cookies by default. This means brands must prepare for a mixed environment: some users will opt into enhanced privacy controls while others won't, requiring flexible measurement strategies rather than waiting for a universal cutoff. [1]

Universal Opt-Out signals are now a legal requirement. Colorado requires controllers to *detect and honor* state-recognized UOOMs; the AG has recognized **Global Privacy Control (GPC),** and enforcement began **July 1, 2024**. Connecticut requires honoring browser-based opt-out signals effective **January 1, 2025**; New Jersey's requirement applies by **July 15, 2025**. Build and test for these signals across all downstream systems.[2] [3] [4] For Northbridge, this meant updating their tag management system to detect GPC (Global Privacy Control) signals, modifying their consent management platform to process opt-outs automatically, and ensuring all third-party vendors

1. Google. "Next steps for Privacy Sandbox and tracking protections in Chrome." April 22, 2025.

2. Universal Opt-Out and the Colorado Privacy Act

3. CT.GOV. Connecticut's Office of the Attorney General. Attorney General Tong advises Connecticut consumers and businesses of opt out rights and requirements. Dec 30, 2024

4. New Jersey Division of Consumer Affairs. Office of the Attorney General.

in their marketing stack could receive and honor these signals within the required timeframes.

California's CPPA (California Privacy Protection Agency) approved new regulations on Automated Decision-Making Technology (ADMT), Risk Assessments, and Cybersecurity Audits on July 24, 2025, with approval by the Office of Administrative Law (OAL) expected by October 1, 2025. Once effective, ADMT compliance will be due on January 1, 2027. Risk assessments for existing processing must be completed by December 31, 2027. Annual attestations will begin on April 1, 2028, and independent cybersecurity audits will be conducted on a staggered schedule starting in 2028.[5] [6] [7] [8]

Each law differs in detail but shares common themes, explicit consent, purpose limitation, data minimization, and consumer rights to access or delete personal information. For a manufacturer like Northbridge, customer trust and legal compliance are now inseparable.

The Consent Economy Principles

Privacy laws create minimum requirements, but customer trust demands maximum effort. Northbridge discovered that simply meeting legal standards wasn't

---

5. CPPA Board Finalizes New Rules on ADMT, Cybersecurity Audits, and Risk Assessments Aug 4, 2025

6. Perkins Coie. CPPA Approves Cybersecurity, Automated Decision Making, and Risk Assessment Regulations. Aug 11, 2025

7. onetrust. CCPA adopts new CCPA regulations: What businesses need to know. Jul 31, 2025

8. Baker Botts. The CPPA Finalizes Rules on ADMT, Risk Assessments, and Cybersecurity Audits. Aug 18, 2025

enough; customers could sense the difference between grudging compliance and genuine respect for their data. The consent economy operates on the principle of reciprocity: customers share valuable personal information in exchange for meaningful value, but only when they believe that the exchange is fair, transparent, and revocable. The following five principles formed the foundation for Northbridge to build data relationships that strengthened over time, rather than eroding through mistrust or neglect.

1. **Transparency** – Explain plainly what data is collected and why.
2. **Control** – Provide easy mechanisms for customers to manage preferences.
3. **Value Exchange** – Offer a tangible benefit for sharing data (rebates, tailored content, loyalty perks). See Chapter 7, "Consumer Insight Collection."
4. **Respect** – Honor revocation promptly; no silent workarounds.
5. **Security** – Safeguard data proportional to its sensitivity.

Executive Summary - Privacy ROI: Privacy compliance isn't a cost center—it's a revenue multiplier. Companies with transparent, consent-based data collection see higher engagement rates, stronger customer relationships, and faster partnership approvals. The upfront investment in privacy infrastructure pays dividends through higher-quality signals and reduced legal risk, turning compliance into a competitive advantage.

Northbridge's Privacy by Design Steps

They committed to embedding privacy guardrails into every touchpoint, whether initiated by Northbridge or its partners:

1. **Unified Consent Ledger** – Data engineering links every identity record

to a consent status table with timestamps and source.

2. **Granular Preferences** – Marketing introduces a preference center that allows customers to select individual newsletters, promotional offers, or product launch alerts, rather than opting for all or nothing.

3. **Revocation Automation** – When a customer opts out, workflows halt emails, flag data for deletion after legal retention requirements, and update downstream systems.

4. **Audit Trails** – Quarterly legal audits to verify consent flags aligned across CRM, ESP, and analytics systems.

5. **Minimal Viable Data** – Forms capture only necessary fields; advanced attributes are added via progressive profiling when the value is proven.

Risk and Opportunity

Neglecting compliance could bring fines and reputational damage, but strong privacy practices offer strategic upside:

- **Trust Differentiator** – Brands seen as transparent and respectful earned higher engagement and loyalty.

- **Signal Quality** – Opt-in contacts are more engaged, resulting in higher conversion and retention rates.

- **Operational Clarity** – Having one authoritative source of consent truth eliminated internal debates and expedited approvals.

Northbridge discovered that opted-in email subscribers showed 3.2 times higher open rates and 4.7 times higher conversion rates compared to legacy lists, demonstrating that consent quality directly impacts campaign performance.

Proactive Compliance: With risk assessments becoming mandatory by 2027, Northbridge began documenting its data processing activities now, creating templates that would satisfy future audit requirements while improving current operations.

Partnership Privacy in Practice

As Jasmine closes the binder, Malik shifts the conversation. "We've been talking about internal data capture. But what happens when that data comes from our partners?"

Northbridge's upcoming cross-promotion with an online recipe platform and a chain of independent coffee shops becomes the case study. Each channel will collect customer interactions:

- **Online** – QR codes embedded in recipe blog posts, along with links in emails, each prompting a flavor preference quiz before entry into a sweepstakes.

- **In-store** – Shelf talkers in coffee shops offering "Scan to win" loyalty perks tied to beverage purchases.

Jasmine points out that partner-collected data must meet the same consent and transparency standards as Northbridge's own. The partnership team drafts joint data-handling guidelines:

- Standardized consent language that appears identically online and in-store.

- Shared consent ledgers ensure Northbridge receives only records with verified opt-ins.

- Partner training sessions to align on revocation protocols and reporting formats.

    - Standardized consent language that appears identically online and in-store.

- Shared consent ledgers, which are digital records maintained and updated by all partners, ensure Northbridge receives only records with verified opt-ins.
- Partner training sessions to align on revocation protocols and reporting formats.
- Data segregation protocols, which involve clear systems to separate and tag partner-sourced data, ensure that this data remains identifiable and can be deleted if necessary..
- Technical specifications for using APIs (software interfaces that enable communication between different computer systems) enable the synchronization of consent information between systems.
- Finally, contractual clauses define liability and audit rights in the event of consent violations.

Evelyn sees the upside: "When our privacy posture is strong, we can activate partner channels faster, without months of legal back-and-forth."

Action Steps Implemented

- Integrated consent flags into the Capture Pulse dashboard, reviewed weekly.
- Added consent decay to KPIs; marketing now plans re-engagement before records lapse.
- Established a privacy incident response playbook with defined roles and timelines.
- Deployed unified consent ledger linking all customer touchpoints to

central preference management.

- Automated Global Privacy Control detection across websites, email, and partner systems.
- Implemented a granular preference center allowing channel-specific opt-ins and opt-outs.
- Created partner consent synchronization protocols with 4-hour propagation requirements.
- Established quarterly cross-system consent audits to verify alignment between CRM, ESP, and analytics.
- Documented all automated decision-making processes to prepare for 2027 ADMT compliance requirements.

Sidebar: Real-world example

Promolytics provides brands with the tools to enable trusted partners, ranging from retailers to event organizers, to collect marketing data on their behalf, with explicit permission prompts built in. Partner-sourced first-party data flows into unified consumer profiles while preserving source attribution, so brands can measure channel-level ROI without sacrificing compliance.

Every consent record is tied to a single profile, making it easy to honor opt-outs instantly and ensure marketing activities comply with state privacy laws. The result is a privacy-by-design ecosystem where brands and partners can collaborate confidently, knowing that data capture strengthens trust rather than jeopardizing it.

Cross-Functional Implementation: Privacy Accountability Matrix

Privacy compliance requires coordination across every function that handles customer data. Unlike traditional legal reviews that happen at project completion, privacy-by-design embeds compliance checkpoints throughout the customer journey. When a marketing campaign collects new data types, legal must pre-approve the consent language, data engineering must update the consent ledger, and customer success must prepare for preference management requests. The following accountability matrix ensures that privacy becomes operational muscle memory, rather than an after-the-fact cleanup.

- **Legal/Compliance:** Owns consent language standards, privacy impact assessments, and regulatory interpretation. Reviews all new data collection methods and partner agreements before launch.

- **Data Engineering:** Manages consent ledger infrastructure, automated opt-out workflows, and data retention policies. Ensures consent status propagates across all systems within required timeframes.

- **Marketing Operations:** Implements preference centers, segments audiences by consent status, and monitors engagement quality of opted-in contacts. Tracks consent decay and plans re-engagement campaigns.

- **Customer Success:** Handles privacy-related customer inquiries, processes data deletion requests, and provides feedback on consent experience friction points.

- **Product Management:** Designs consent flows that balance legal requirements with user experience, conducts A/B testing on consent conversion rates, and prioritizes features that enhance privacy.

Executive Reflection – Compliance Advantage:

Looking at your current data collection touchpoints—website forms, email subscriptions, partner integrations, and promotional campaigns—which generates the least transparent customer experience? How can you redesign that touchpoint to clearly communicate data use, provide granular control options, and demonstrate immediate value exchange while satisfying state privacy requirements, such as Global Privacy Control and automated decision-making disclosures?

Chapter 11 Checklist: Embedding Privacy by Design

- Map all active data collection points and confirm that visible consent language is present.
- Centralize consent status into a single ledger accessible by all systems.
- Automate revocation workflows to propagate opt-outs within 24 hours.
- Implement Global Privacy Control detection across all customer touchpoints.
- Deploy a granular preference center with channel-specific opt-in controls.
- Establish partner consent synchronization protocols with defined SLAs.
- Complete partner training on consent standards and revocation procedures.
- Document automated decision-making processes for ADMT (Automated Decision-Making Technology) compliance preparation.
- Schedule quarterly consent alignment audits across systems.

- Add consent decay and opt-out trends to the executive KPI dashboard.
- Create a privacy incident response playbook that defines cross-functional roles.

By embedding privacy into every touchpoint, Northbridge not only avoided compliance risks but also increased the value of its marketing data. As a result, every permissioned contact in the database represented a real, willing participant in the brand's story. That trust shifted the focus from "How do we reach more people?" to "How do we engage the right people in the right way?"

The privacy infrastructure Northbridge built didn't just ensure compliance; it enhanced their AI capabilities, as discussed in Chapter 10. With clean, consented data flowing through documented pipelines, their models could train on high-quality signals while maintaining audit trails for any automated decisions, satisfying both current requirements and preparing for ADMT regulations ahead.

With this consent foundation firmly established, Northbridge's marketing team now turns to transforming permissioned data into actionable customer intelligence. The next chapter examines how Northbridge converts its clean, consented data streams into behavioral segments that reveal purchase patters, predict response likelihood, and identify high-value customer who drive measurable growth rather than relying on demographic assumptions that sound logical but fail to convert.

Chapter Twelve

Customer Truth

Turning First-Party Data Into Segments, Decisions, & Measurable Growth

First-party data reveals what customers actually do, versus what they claim to do. When organizations directly capture permission-based behavioral signals, they gain insight into purchase patterns, channel preferences, price sensitivity, and response timing. Third-party sources can only estimate these details. This reality supports segmentation strategies that fuel measurable growth, rather than theoretical personas that fail to convert.

Leaders with strong first-party data can answer key questions for competitive positioning: Which customers generate 80% of profit? What encourages lapsed customers to return? Which promotions drive trial or repeat purchases? How do regional preferences impact national rollouts? When do seasonal buyers become year-round customers? These answers are based on real behaviors, not aggregated customer averages.

This chapter now shifts to providing frameworks for transforming raw first-party data into actionable segments, along with testing protocols that isolate causation from correlation, and measurement systems that connect customer actions to financial outcomes. Remaining practical rather than theoretical, Northbridge's experience demonstrates how systematic segment development can accelerate revenue growth when grounded in verified customer behavior rather than demographic assumptions.

Northbridge Challenge

Chris Chen, Chief Revenue Officer at Northbridge, had a clear charge: increase customer lifetime value by 20% this fiscal year while maintaining margins. The timeline was tight, with a board presentation due in six months and full implementation by year-end. Traditional mass retail promotions had diminished returns: trad spend consumed 18% of gross revenue while acquisitions costs rose.

The available inputs were both promising and complex. Northbridge amassed 1.2million opted-in emails, 340,000 SMS-consented mobile numbers, 890,000 loyalty members with varied engagement levels, and 2.3 million validated purchases over a 24-month period. Promotions recently tested various mechanics: percent discounts, dollar-off thresholds, BOGO offers, flash sales, unique customer experiences, and points multipliers. Consent was strong, with 92% of know customers percent of customers permitting at least one channel and 46% opting into multiple channels.

Constraints shaped the approach much as opportunities. The initiatives budget was capped at $3.2 million, covering technology, analytics, and execution. The channel mix had to include existing retail partners, which drove 65% of the revenue. Key partners required 90-day promotional calendars, limiting agility. Legal mandates included strict consent checks, data retention limits, and age-gating for select products. These boundaries forced disciplined focus on high-impact first-party data efforts.

Data Inventory Checklist

Identity resolution, the process of linking multiple data points to a single customer, forms the foundation of customer truth. Organizations must map identifies across touchpoints. These include email addresses, mobile numbers, loyalty IDs, and device IDs. Permission tracking requires granular records. These include: opt-in date, channel, method (explicit or inferred), frequency preferences, and unsubscribe history. Event types span the customer journey: user experiences, website visits, email opens, SMS clicks, app sessions, store visits (via geolocation or provided directly from the customer), and social engagement. Product mapping connects SKUs to customer preferences, including category affinity, price sensitivity, size, flavor profiles, and purchase occasions. Store mapping links customers to retail environments. It includes primary stores, secondary locations, channel preferences (in-store, pickup, delivery) and proximity to competitors. Receipt verification closes the loop. It includes the purchase date, location, item, price paid, promotion redeemed, and payment method category.

Profile Schema Highlights

Required fields create minimum profiles. Device ID, email, or mobile as the unique identifier. Consent status must be a timestamped. First purchase date sets lifecycle stage. Last purchase date prompts reactivation. Total purchases show relationship depth. Geographic fields enable location targeting: Zip codes for mail, DMAs (Nielsen-defined media markets) for planning, and store proximity for retail.

Optional fields enhance targeting precision when available. Demographic appends (additional information such as age range and household composition) inform creative messaging. Channel engagement scores (rating of interaction via email, SMS, and social media) optimize contact strategy. Product preferences derived from purchase history guide recommendation engines. Promotional responsiveness indicates offer sensitivity. Customer service interactions flag satisfaction issues. Social sharing behavior identifies advocates. Survey responses add depth in psychographics (attitudes, lifestyle, interests). App usage patterns reveal engagement. Weather sensitivity links purchase timing with environmental factors, including temperature and rainfall.

Segmentation Templates

Lifecycle stages follow predictable patterns that drive different strategic responses:

Stage	Definition	Typical Size	Primary Objective	Recommended Tactics
New	First purchase within 90 days	15-20%	Drive second purchase	Welcome series, category education, trial incentives
Active	Purchase within last 6 months	35-40%	Increase frequency	Cross-sell, loyalty rewards, exclusive access
At-Risk	No purchase in 6-12 months	25-30%	Prevent churn	Win-back offers, feedback surveys, channel reactivation
Lapsed	No purchase in 12+ months	20-25%	Reactivate or remove	Aggressive offers, permission refresh, list hygiene

Lifecycle Stage Tactics

RFM (Recency, Frequency, Monetary) segmentation ranks customers based on their transaction behavior to identify who deserves more investment versus who needs different strategies. This proven framework combines three dimensions of purchase history into actionable segments:

RFM Dimension	Band/Tier	Definition	Typical Size	Primary Strategy
Recency				
Hot	0-30 days	Purchased last month	8-12%	Capitalize on engagement
Warm	31-90 days	Purchased last quarter	15-20%	Maintain momentum
Cooling	91-180 days	Purchased last 6 months	20-25%	Reactivate interest
Cold	181-365 days	Purchased last year	25-30%	Win-back campaigns
Frozen	365+ days	Over a year inactive	20-25%	Aggressive offers or remove
Frequency				
Trial	1 purchase	Single transaction only	35-40%	Drive second purchase
Emerging	2-3 purchases	Early repeat behavior	25-30%	Build habit
Established	4-6 purchases	Regular customer	20-25%	Increase basket size
Loyal	7+ purchases	High frequency buyer	10-15%	VIP treatment
Monetary				
Price-Sensitive	Bottom tercile	Low average order value	33%	Value messaging
Balanced	Middle tercile	Average spending	34%	Mix of value and premium
Premium	Top tercile	High spenders	33%	Quality and exclusive offers

Recency, Frequency, Monetary (RFM)

Propensity scores use predictive analytics to identify which customers are most likely to take specific actions, enabling efficient resource allocation by focusing efforts on those with highest probability of response:

Propensity Level	Model Score	Population	Response vs Baseline	Recommended Action
High	>0.8	Top 20%	3x baseline	Maximum investment, personalized offers, multi-channel contact
Medium	0.3-0.8	Middle 50%	0.8-1.2x baseline	Standard campaigns, test new approaches, monitor for movement
Low	<0.3	Bottom 30%	0.3x baseline	Suppress from expensive channels, research barriers, different strategy

Propensity Modeling

Occasion-based segments recognize that customer needs vary by context, requiring different messages and offers depending on the purchase situation:

Occasion Type	Identifying Signals	Typical Size	Message Focus	Optimal Timing
Seasonal Shoppers	Purchase clusters in specific periods	25-30%	Holiday themes, gift guides, party planning	3-4 weeks before occasion
Event-Triggered	Purchases near personal dates	15-20%	Celebration, milestones, personal recognition	1-2 weeks before known dates
Stock-Up Buyers	Large quantity, low frequency	20-25%	Bulk savings, storage tips, subscription offers	During promotional windows
Immediate Consumption	Small quantity, high frequency	30-35%	Convenience, freshness, instant gratification	Continuous availability
Gift Purchasers	Different ship-to address, gift options	10-15%	Gift wrap, recommendations, recipient preferences	Peak gift seasons

Occasion Segments

Northbridge Implementation: From Data to Revenue

Northbridge's customer truth initiative began with data preparation and analysis. The analytics team identified an addressable universe of 940,000 unique customers with verified consent status. The team built 12 initial segments based on lifecycle stage (4 groups), purchase frequency (3 tiers), and category preference (5 clusters), creating manageable test cells of 15,000 to 80,000 customers each.

Offer design followed a test-and-learn philosophy across three waves. Wave 1 tested discount depths: 15% off, 20% off, and $5 off a $25 purchase. The high-frequency segment showed no incremental response above 15%, whereas lapsed customers required a 20% increase to reactivate. Wave 2 tested mechanics: straight discount versus BOGO versus points multiplier. Category buyers preferred BOGO for sock-up behavior, while variety seekers responded to points for flexibility. Wave 3 tested channels: email only versus email plus SMS versus full omnichannel. Multi-channel contact increased response by 2.3x but required frequency capping to prevent fatigue.

Launch execution required careful orchestration across channels. Email campaigns are deployed on Tuesdays and Thursdays at 10:00 AM local time, based on engagement analysis. SMS messages followed by email within 4 hours for non-openers. Social audiences are refreshed daily with 7-day conversion windows. Geo-fences are activated during peak shopping hours (Thursday through Sunday). The monitoring dashboard tracked hourly response rates, flagging any segment that performed below 80% of the forecast for review.

Results exceeded targets within the first quarter. The purchase rate among active customers increased from 24% to 31% (a 29% lift). The reactivation rate for lapsed customers jumped from 8% to 14% (a 75% increase). The new customer second-purchase rate improved from 35% to 47% (a 34% lift). Store velocity for featured SKUs increased 18% in high-treatment DMAs. The combined impact delivered $4.7 million in incremental revenue on a $890,000 investment, resulting in a 5.3x ROI.

Chris's was impressed with the way the cross-functional team worked together to coordinate the required weekly standups with defined workstreams. Marketing owned segment definitions, creative development, and campaign execution. Sales managed retailer partner communication and field activation. Finance tracked investment, validated lift calculations, and reported ROI. Legal reviewed consent protocols, data handling, and promotional terms. Data & Analytics built segments, monitored performance, and produced insights. Each function had documented responsibilities, deadlines, and escalations paths in place.

Sidebar: Real World Example

Promolytics manages consent tracking with per-channel permission status, time and date stamps, and source, so the team can filter eligible audiences within the system and remain compliant. Unified profiles stitch QR scans, surveys, coupon events, and receipt information into a single person unique ID. The segment builder provides the team with the ability to filter information down to lifecycle, RFM, and propensity buckets. Dashboards show purchase rate, lift, and store-level velocity. The result is fewer manual steps and shorter time from signal to decision.

Action Steps Implemented

Northbridge executed customer truth initiatives through documented phases with clear ownership:

Phase 1: Data Foundation

- Data audit completed by Analytics team
- Identity resolution rules defined by Data Governance
- Consent verification process documented by Legal
- Profile schema approved by Marketing Leadership
- Historical baseline metrics established by Finance

Phase 2: Segmentation Development

- Lifecycle segments created: New (112,000), Active (347,000), At-Risk

(293,000), Lapsed (188,000)

- RFM scoring applied across all customers with purchase history
- Propensity models trained on 18 months historical data
- Test/control groups randomly assigned with 80/20 split
- Segment performance forecasts documented and approved

Phase 3: Campaign Execution

Segment	Offer Tested	Channel Mix	Test Size	Control Size	Primary KPI
New Customers	20% off second purchase	Email + SMS	89,600	22,400	Second purchase rate
Active High-Value	Early access + 15% off	Email + Social	52,800	13,200	Average order value
At-Risk	$10 off $40	Email + SMS + Geo	234,400	58,600	Reactivation rate
Lapsed	25% off + free shipping	Email + Social	150,400	37,600	Win-back rate

Northbridge Campaign Information

Phase 4: Results Analysis

- Weekly performance reports produced every Monday
- Mid-flight optimization decisions documented in cross-functional team meeting minutes
- Final ROI calculation validated by Finance
- Executive presentation delivered to board
- Playbook documentation completed for scaling

Cross-Functional Implementation

Customer truth initiatives require systematic accountability across departments to prevent data gaps and ensure reliable insights. Each team must own specific aspects of segment integrity based on their access to data sources and decision-making authority.

- **Marketing:** Implement consent tracking, set segment size thresholds, customer feedback data, and integrate cross-channel engagement data
- **Sales:** Provide real-time retail performance
- **Finance:** Enforce minimum statistical requirements before budget allocation decisions
- **Operations**: Alert teams when promotional demand exceeds inventory availability
- **Analytics**: Run weekly data quality audits and segment performance checks

Executive Reflection:

How do your current customer segments drive different business outcomes, and what evidence validates these claims? If the evidence is thin or anecdotal, you will need to treat the claims as hypotheses and redesign the test.

Now ask: how could first-party customer information raise the game for your team by sharpening segment definitions, improving consented reach, enabling targeted offers, and closing the loop with verified outcomes.

Chapter 12 Checklist: Segmentation and Measurement

- Data Readiness: unified profiles, consent fields, product and store mappings.

- Segment definitions approved and documented.
- Test design complete with baselines, holdouts, and pivot rules.
- Launch criteria and guardrails set and shared.
- Read-out schedule on the calendar with owners.
- Archive artifacts and learnings in a shared repository.

Customer truth initiatives generate validated segments and behavioral insights that transform marketing from broad awareness campaigns to precision activation.

The next chapter explores how these segments enable sophisticated channel strategies that optimize message, timing, and creative for each customer micro-moment. Marketing teams that master customer truth acceleration achieve improved ROI by eliminating waste on customers who won't respond while intensifying investment in those demonstrating purchasing readiness. The frameworks for campaign orchestration build directly on the segmentation and measurement foundations established here, turning customer understanding into revenue acceleration.

Chapter Thirteen

First-Party Powered Marketing:

When Precision Targeting Create Measurable Business Results

Marketing execution transforms from creative guesswork into scientific measurement through first-party audience design, message personalization, and incremental lift validation. This chapter provides the precision marketing framework that turns consented customer signals into verified revenue impact through coordinated campaign optimization and sales enablement.

We'll examine five filtering techniques that refine raw first-party data into market-ready microsegments, introduce dynamic audience models that respond to real-time behavioral signals, and provide campaign frameworks that ensure consistent insight-to-action velocity; the speed at which marketing insights translate into measurable customer actions. You'll learn to establish measurement protocols that track incremental lift – the verified revenue impact attributable to specific marketing interventions – and optimization systems across the entire campaign lifecycle.

By the end of this chapter, you'll have operational blueprints for implementing precision marketing process, audience segmentation frameworks that maintain relevance while scaling personalization, and measurement systems that prove marketing ROI while accelerating continuous campaign improvement.

From Broadcast to Precision

Traditional mass campaigns cast broad nets, hoping to attract some interest. Now, contacts in Northbridge's database are opted-in and linked to verified purchase activity where relevant. The team can replace assumptions with accuracy. First-party data transforms marketing from broad outreach into targeted, high-value engagement.

Why this shift matters:

Before focusing on tactics, consider how a compliant, insight-rich approach strategically distances your team from legacy models. The following comparison illustrates how first-party data revolutionizes every aspect of marketing execution, transitioning from assumption-based tactics to evidence-driven precision. As you review, identify areas where your marketing aligns with the "Old Approach," and prioritize which row you will address first in your next campaign.

Old Approach	New Approach with First Party Insight
Broad demographic targeting purchased from a media broker	Cohort targeting based on first-party consumer data and receipt-verified purchase behavior
One creative, one offer for all	Dynamic content matched to lifecycle stage and preference
Success is measured by impressions and CTR	Success is measured by incremental lift in verified sales
Budget fixed upfront	Budget is reallocated periodically based on real-time performance

From Broadcast to Precision

Filtering First-Party Signals: From Firehose to Focus

First-party data is powerful because it's dense: receipts, scans, clicks, preferences, and timing all land on a single profile. But density without discipline becomes noise. The job is to filter signals so that only decision-grade audiences, messages, and moments reach the market. Use the framework below to go from *everything we know* to *what we'll act on,* and prove it with lift.

The Five Filters

1. **Permission** — Start with consent. Include only opted-in profiles with current permission and valid contact points.

2. **Integrity** — Exclude bad reads. Remove bot traffic, duplicate receipts, test scans, MPP-only opens (Apple's Mail Privacy Protection, which artificially inflates open rates by pre-loading email content), and stale attributes.

3. **Relevance** — Narrow to people whose **recent** behavior maps to the objective (trial, repeat, trade-up, bundle, win-back).

4. **Readiness** — Detect buying windows and constraints (daypart, pay cycle, store inventory, weather, proximity).
   Example: Convenience shoppers who clicked between 2–5 pm and purchased within 24 hours the last time.

5. **Reachability** — Choose the channel they actually respond to (SMS vs. email vs. social media), with fatigue and frequency caps.
   Guardrail: pause if engagement drops below your fatigue threshold (set this in your "Key Metrics That Matter").

   - Example fatigue thresholds:

     - Email: Pause if open rate drops below 15% or click rate below 2%

     - SMS: Pause if the opt-out rate exceeds 1% weekly or the response rate falls below 5%

     - Overall: No more than 3 touchpoints per week across all channels

When all five filters are true, you've got a **market-ready micro-segment**: permissioned, clean, relevant, primed, and reachable.

Executive Summary - Precision Filtering: The five-filter system transforms broad first-party data into market-ready micro-segments that actually convert. Permission ensures legal compliance, Integrity removes bad data, Relevance targets behavior-matched prospects, Readiness identifies buying windows, and Reachability selects optimal channels. This disciplined approach typically improves conversion rates 2-3x while reducing cost per acquisition, proving that precision beats volume in first-party marketing.

Focus Recipe in Action: The Tuesday Morning Sprint

Tuesday, 10:47 AM. Marketing analyst Sarah Chen notices an alert on her dashboard: QR code scans in the Dallas-Fort Worth metro have jumped 340% over the past 72 hours. The Regional Surge recipe triggers automatically.

"We've got a live one," Sarah announces to the team clustered around the standing desk. The Capture Pulse dashboard shows the spike started Sunday evening and accelerated through Monday, with scan activity concentrated in three ZIP codes around Plano and Richardson.

Malik pulls up the inventory feed. "Good news—our DFW distribution partner shows 85% stock levels across target stores. We're above the 70% threshold."

Evelyn checks the consent ledger. "How many permissioned contacts do we have in those ZIPs?"

"2,347 profiles with SMS consent, 1,892 showing 'Explorer' behavior patterns," Sarah reports, applying the Relevance filter. "They've engaged with content but haven't converted yet."

The team runs through the remaining filters in sequence. Permission: verified. Integrity: bot traffic removed, duplicate scans filtered out. Readiness: the spike timing coincides with local food blogger mentions and a competitor's supply shortage. Reachability: SMS engagement in this metro averages 23% during store hours.

"That gets us to 1,204 market-ready contacts," Sarah confirms. "Above our 1,000 minimum."

Jasmine, monitoring from legal, gives the compliance thumbs-up. "All contacts opted in within the past 90 days. GPC signals processed. We're clear to proceed."

The creative team has a templated message ready: "In stock near you—try it this week" with store locator deep-links. The offer is a modest 48-hour rebate co-funded by their retail partner.

"Launch at 1 PM when stores hit peak foot traffic," Evelyn decides. "SMS only, four-hour window, throttle sends based on real-time inventory updates."

By 1:15 PM, the campaign is live. The team monitors real-time metrics: a 31% open rate, an 8% click-through rate to the store locator, and early redemption signals are starting to appear in the dashboard.

"Store manager in Richardson just texted," Malik reports, reading from his phone. "They've had six people mention the SMS in the past hour. Shelf velocity is definitely up."

Sarah sets the auto-pause triggers: if inventory drops below 40% or if redemption rate exceeds projections by 200%, the campaign stops automatically.

Wednesday morning brings the results. The test generated 127 incremental store visits and 89 verified purchases across the target ZIP codes—a 7.2% conversion rate that's 2.3 times their baseline. Cost per incremental unit: $4.80, well below their $8 target.

"Regional Surge recipe validated," Evelyn notes in the campaign log. "Template updated with DFW parameters. Ready for the next spike."

The entire cycle—from signal detection to validated results—took 28 hours. No committee meetings, no approval chains, just disciplined execution of a proven framework with real-time guardrails and immediate measurement.

Sarah updates the Focus Recipe library with new insights: the DFW metro responds best to afternoon timing, store locator functionality drives higher conversion rates than generic offers, and co-op funding improves both retailer relationships and unit economics.

The playbook gets sharper with each execution.

"Focus Recipes": Scenarios You Can Run

Each recipe follows the same structure: **Aim → Filters → Segment → Message → Channel/Timing → Offer → Measurement,** so teams can build consistently and scale what works.

1) Repeat Lift: Turn Trial into Habit (30-day window)

- **Aim:** Increase the rate of second purchases among recent trialists.
- **Filters:**
    - Permissioned; receipt-verified **first** purchase in past 7–21 days
    - No second purchase yet; no opt-out; fatigue < threshold
    - Integrity: exclude coupon abusers (≥3 redemptions/month)
- **Segment:** "High-intent trialists" (clicked recipe content OR rated product ≥4/5)
- **Message:** "Keep it in your routine—our quick 2-minute 'best pairing' guide"
- **Channel/Timing:** SMS at 3–5pm local (historic convert window)
- **Offer:** Light nudge (recipe + modest add-on rebate), not a deep discount
- **Measurement:** Matched holdout; Overall Evaluation Criterion (OEC = 14-day repeat rate; guardrails = unsubscribe + margin

2) Cross-Sell: Bundle the Next Best Product

- **Aim:** Raise basket value for customers who repurchased once.
- **Filters:**
    - Permissioned; repeat in the last 60 days; no prior bundle purchase
    - Integrity: POS match present; stable ID; no conflicting consent
- **Segment:** "Citrus fans" who also viewed salty snacks content
- **Message:** "Your citrus go-to, now with the perfect salty pair"
- **Channel/Timing:** Email in the morning; reminder SMS next afternoon if unopened
- **Offer:** Bundle at a price fence that preserves margin (learned from prior elasticity tests)
- **Measurement:** Incremental basket lift vs. holdout; track repeat at 30 days

3) Regional Surge: Capture Local Demand Spikes

- **Aim:** Convert a sudden interest spike into verified sales before it fades.
- **Filters:**
    - Permissioned; region = metro cluster with ↑ scans/visits past 72 hours
    - Integrity: exclude bot-like scans; confirm inventory ≥ threshold

- **Segment:** "Explorers" (engaged but not yet purchased) in those ZIPs
- **Message:** "In stock near you—try it this week" + store locator deep-link
- **Channel/Timing:** Push/SMS during store open hours; throttle by inventory feed
- **Offer:** Small, time-boxed rebate; co-op with retailer if possible
- **Measurement:** Store-level incremental units vs. matched stores; end after 7 days

4) Churn Rescue: Win Back Quiet Loyalists

- **Aim:** Re-engage high-value buyers who went quiet.
- **Filters:**
    - Permissioned; LTV in top 25%; no purchase in 60–90 days
    - Integrity: remove contacts with hard-bounces or prior suppression
- **Segment:** "Loyal switch-risks" (recent competitor view signals or survey mentions)
- **Message:** Value reinforcement (taste claims, quality proof) + "What would bring you back?" one-tap poll
- **Channel/Timing:** Email first (rich content), follow with support-style SMS if poll = "had an issue"
- **Offer:** Choose between loyalty credit **or** early access (avoid defaulting to discounts)
- **Measurement:** Win-back rate vs. holdout; track cost per incremental

reactivation

5) Price-Sensitive Cohort: Hold Margin, Still Convert

- **Aim:** Convert bargain-leaning shoppers without training the whole base to expect discounts.
- **Filters:**
    - Permissioned; ≥2 redemptions in 60 days; cart abandons at small price increases
    - Segment: "Value seekers" only
- **Message:** "Best value this week: [bundle or larger pack]"
- **Channel/Timing:** Email on pay-cycle days; SMS right before typical purchase time
- **Offer:** Bundle or pack-size value—**not** a deeper coupon on a single unit
- **Measurement:** Margin per incremental unit vs. single-unit discount baseline

How to Build the Segment (one-page brief your team can reuse)

- **Objective & OEC:** e.g., "Boost repeat @14 days; measure with matched holdout."
- **Inclusion rules:** consent = true; last_event ≤ X days; verified purchase = true; channel reachable = SMS|Email.
- **Exclusion rules:** fatigue flag; bot flag; coupon abuse; inventory < threshold; "do not disturb" hours.

- **Minimums:** ≥1,000 contacts *or* 90 days of data per region before scale.
- **Creative/messaging:** one insight-led angle; ≤2 variants per cohort.
- **Offer policy:** price fence; cap redemptions; sunset date.
- **Holdout sizing:** 5-10% for large segments (>10,000), 10-20% for smaller segments to ensure statistical significance
- **Ops & risk:** SLA to pause if stock-outs or unsubscribes breach guardrails.
- **Readout:** lift vs. holdout; cost per incremental unit; decision (scale/stop/tweak).

Do / Don't (to keep signal clean)

- **Do** pre-register success metrics and guardrails with the Finance team.
- **Do** suppress MPP-only opens; optimize for clicks, replies, and **receipt-verified** outcomes.
- **Do** cap frequency and rotate channels to avoid fatigue and list decay.
- **Don't** mix unverified third-party segments into first-party reads; keep tests apples-to-apples.
- **Don't** scale a tactic before you have a **minimum sample** and a stable read.
- **Don't** leave winners unstandardized, publish a **Lift Card** (a one page activation, showing test design, control group, and cost per incremental unit), and templatize. See example: Appendix H.

Scale Considerations: Right-Sizing Your Approach

Not every organization needs every filter or recipe from day one. Your scale determines where to focus first and how sophisticated your segmentation should be. Here's how to calibrate your approach based on database size and resources:

Startup Stage (<10,000 customers)

- Focus on mastering 2-3 core segments maximum (e.g., "new customers," "repeat buyers," "at-risk")
- Prioritize the Permission and Integrity filters, get the basics right before adding complexity
- Run one Focus Recipe at a time with manual analysis
- Use simple A/B tests rather than complex holdouts
- Key win: Prove you can measure incremental lift on one segment before expanding

Growth Stage (10,000-100,000 customers)

- Deploy all five filters, but keep Readiness and Reachability rules simple
- Implement 3-5 Focus Recipes running in parallel
- Invest in automation for segment updates (daily or weekly refresh)
- Maintain 10% holdouts for statistical significance
- Add predictive scores for high-value segments only
- Key win: Standardize your top 3 performing recipes into repeatable playbooks

Enterprise Stage (>100,000 customers)

- Apply all five filters with ML-driven scoring for Relevance and Readiness
- Run 10+ concurrent micro-segments with automated recipe selection
- Implement real-time segment updates triggered by behavior
- Use sophisticated holdout strategies (stratified sampling, synthetic controls)
- Deploy AI for creative personalization within segments
- Key win: Achieve segment-of-one personalization for the top 20% of customers

Resource Reality Check:

- Need a minimum of 1,000 contacts per segment for reliable measurement
- Require 30-60 days of data before declaring a recipe successful
- Must have at least 2x the expected conversions in both test and control for significance
- If you can't properly measure it, don't scale it; better to run fewer recipes well than many poorly

Northbridge Mini-Examples (how the filters changed the work)

- **Afternoon SMS > Evening Email:** After filtering for daypart readiness and consenting SMS users, the repeat lift rose by +11% compared to email-only; fatigue remained below the threshold thanks to a weekly cap.
- **Bundle > Deeper Discount:** Filtering out chronic coupon users and targeting value seekers, a bundle raised its **margin per incremental unit** by **+8%**
- **Local Spike Capture:** Applying integrity (bot removal) + inventory checks prevented wasted media; verified that store sales in the hot ZIP codes rose by **double-digits** week-over-week.

Four Marketing Levers Powered by Insight

First-party signals turn strategy into levers you can actually pull. Use these four levers—who you target, what you say, when you show up, and where budget flows—to convert insight into OEC-aligned outcomes. Change one lever at a time, measure with holdouts, and maintain fatigue and consent guardrails.

1. **Audience Design** – Dynamic segments built from lifecycle, preference, and predictive scores. Example: *"High-intent trialists"* receive focused onboarding sequences.
2. **Message Relevance** – Content is tailored to motivations surfaced in sentiment analysis. Sugar-conscious Explorers get nutritional comparisons. Value seekers get bundle offers.
3. **Channel Timing** – Contextual data reveals that afternoon SMS outperforms evening email for convenience shoppers. Cadence adjusted accordingly.

4. **Budget Allocation** – Capture Pulse reveals that urban QR redemptions drive higher incremental lift, shifting spend from national TV to in-store activations.

Channel Conflict Resolution: When Filters Point in Different Directions

Your filters won't always align perfectly. A customer might exhibit high SMS readiness based on past behavior but have explicitly set email as their preferred method. Another option might be reachable via push notifications, but only during hours when email performs more effectively. Here's your decision hierarchy when filters conflict:

Priority Order for Channel Selection:

1. Explicit Consent Trumps Everything

- If the customer selected "email only" in preferences, honor it regardless of SMS performance data
- Never override a channel opt-out, even if other signals suggest receptiveness

2. Recency Beats History

- The last 30 days of engagement outweigh the 90-day averages
- If SMS engagement dropped off recently but email picked up, pivot to email
- Example: Customer historically preferred SMS but opened 3 emails this week → try email first

3. Message-Channel Fit

- Complex content (recipes, reviews) → Email

- Urgent/time-sensitive (flash sale, back in stock) → SMS or Push
- Visual-heavy (new product showcase) → Email or In-app
- Quick confirmations or reminders → SMS

4. Cost-Performance Balance

- Calculate the cost per incremental conversion by channel
- If Email delivers 80% of SMS performance at 10% of the cost, choose email
- Reserve premium channels (SMS, direct mail) for high-value segments or moments

Conflict Resolution Examples:

Scenario 1: High-value customer shows SMS readiness at 3pm but has an email preference set

- **Solution**: Send email at 3pm (honor preference, but use optimal timing)

Scenario 2: Relevance filter says "ready to buy," but Reachability shows fatigue across all channels

- **Solution**: Wait 48-72 hours, then reach out via their least-recently-used permitted channel

Scenario 3: Regional surge opportunity, but the inventory filter shows only 50% availability

- Solution: Geo-target only ZIP codes with confirmed inventory, suppress others despite readiness

The Override Rule:

Allow manual override only for:

- Customer service recovery situations
- Regulatory or safety notifications
- Explicit customer requests ("call me about this order")

Document every override for both compliance and learning purposes.

Northbridge Campaign Case: Autumn Bundle Push

This campaign shows the first-party playbook in action: a clear objective, targeted audiences, tailored creative, and a built-in holdout to prove causality. Designed around incremental repeat purchase as the OEC, the team personalized offers by flavor preference and measured lift against a matched control. The result isn't a story about clicks—it's verified revenue impact at a lower cost per incremental unit.

- **Objective:** Increase repeat purchase among recent first-time buyers.
- **Tactic:** Email and SMS highlighting a limited bundle deal, creative personalized by flavor preference.
- **Holdout:** 10% of eligible customers withheld for lift measurement.
- **Result:** The holdout uplift shows a 9% higher repeat purchase rate among the targeted cohort; the cost per incremental unit is 25% lower than in the prior seasonal campaign.

Sidebar – Real-world example

Promolytics ties coupon redemptions and receipt-verified purchases to each unique customer profile, so you can attribute **incremental sales** to specific messages, creatives, and channels—not to averages. It unifies QR scans, receipt uploads, surveys, and link clicks in one view, letting your team see what's working now and what to change next. The result: in-campaign optimizations and next-cycle plans grounded in first-party proof, not guesswork.

Looking ahead, AI and machine learning will dramatically expand these capabilities: imagine real-time creative adjustments based on emerging audience patterns, dynamic budget shifts triggered by live sell-through data, or predictive scoring that automatically re-prioritizes audience segments mid-campaign. By combining today's actionable analytics with tomorrow's adaptive intelligence, brands can keep every activation both accountable and future-ready.

Key Metrics That Matter

Northbridge replaced vanity metrics with four monthly metrics that demonstrate direct links between marketing and revenue, providing numbers that prove an impact on revenue and loyalty. If a dashboard metric doesn't tie to confirmed revenue or retention, consider removing it.

Metric	Why It Replaces Vanity	Northbridge Target
Verified incremental lift	Moves beyond clicks to revenue impact	≥ 5% per major campaign
Time to first repeat purchase	Indicator of loyalty building	Reduce from 45 to 30 days
Engagement fatigue index	Prevents over-messaging	Keep the opt-out rate under 1% monthly
Conversion by segment	Reveals cohort performance variance	Identify and scale top quartile segments

Key Marketing to Revenue Metrics

Action Steps Implemented

- Replaced static persona decks with live dynamic segments accessible in the marketing platform.
- Embedded incremental lift calculation in campaign briefing template.
- Set periodic budget reallocation review using Track data.
- Introduced a creative variant library mapped to sentiment themes.
- Added fatigue safeguard rule: pause messaging when interaction drops below threshold.

Cross-Functional Implementation: Precision Marketing Accountability Matrix

Precision marketing requires coordination across every function that touches customer experience and revenue measurement. Unlike traditional campaign execution that operates in marketing silos, first-party precision embeds accountability checkpoints throughout the customer journey. Dynamic segments that show buying readiness create a cascade of coordinated activities. Marketing develops relevant messaging, sales prepares qualified outreach materials, and finance establishes incremental tracking. The accountability matrix below ensures this coordination happens consistently rather than sporadically.

- **Marketing Operations:** Owns audience segmentation, message personalization, and campaign performance measurement. Maintains dynamic segments and fatigue thresholds across all channels.
- **Sales Enablement:** Translates marketing insights into account-specific materials, manages handoff processes, and provides field feedback on

customer readiness accuracy.

- **Data Analytics:** Develops predictive scoring models, maintains holdout test integrity, and calculates incremental lift attribution across campaigns and channels.
- **Finance:** Validates incremental revenue calculations, approves budget reallocation protocols, and tracks cost per incremental unit across marketing investments.
- **Customer Success:** Monitors post-campaign customer satisfaction, processes preference updates, and provides feedback on message relevance and timing effectiveness.

Executive Reflection – Revenue Accountability:

If marketing spend were cut by 20% tomorrow, which of the five filtering criteria—Permission, Integrity, Relevance, Readiness, or Reachability—would you double down on to maintain revenue growth? Would you prioritize reaching more permissioned contacts with basic messaging, or focus your reduced budget on the most precisely filtered micro-segments using the Focus Recipes framework? How would this decision change your approach to incremental lift measurement and the marketing-to-sales handoff process?

Chapter 13 Checklist: Turning Insight into Impact

- Build one dynamic segment using predictive scores and behavioral data to improve targeting effectiveness.
- Design campaign objectives around incremental lift, not just reach.
- Implement a holdout test for every central activation to quantify incremental lift.

- Schedule a periodic budget shift meeting based on real-time performance to maximize campaign results.
- Monitor engagement fatigue and automate pause rules to protect list health and sustain long-term engagement.

The Marketing-to-Sales Handoff Kit

Precision marketing campaigns generate more than improved metrics; they create field intelligence that sales teams can use to close deals. When marketing proves that specific messages resonate with particular customer segments, or demonstrates that certain regions show higher purchase intent, that insight becomes a competitive advantage in sales conversations. The most successful organizations don't let these insights die in marketing dashboards; they package them into sales-ready materials that turn campaign learnings into revenue conversations. Northbridge developed a systematic handoff process that ensures every campaign cycle produces actionable intelligence for the field. (See example: Appendix H: Marketing → Sales Handoff Template):

- **Store Heatmap:** Ranks locations by the latest lift in customer interest, linking directly to visit and sales data.
- **Account Briefs:** A summary of target audience segments, most effective marketing messages, and the top-performing incentives driving shopper behavior for the retailer.
- **Lift Cards:** One page per activation, showing test design, control group, and cost per incremental unit.
- **Execution Notes:** Concise documentation of what performed well or poorly in display execution, supplemented by compliance photos.
- **Risk Flags:** Notifications including campaign fatigue thresholds, stock-out warnings, and relevant data quality issues to address with

buyers.

The Cadence That Makes It Real

- **Monday**: Marketing completes analysis and publishes handoff kit, tags accounts in the CRM, and prepares for a sync to align talking points, objections, and next best offers
- **Tuesday 2:00 p.m.:** Marketing-Sales alignment meeting (30 minutes).
- **Wednesday-Thursday**: Sales reps customize materials for specific accounts.
- **Customer meeting:** Sales leads with local evidence, not anecdotes, and asks for specific expansion actions.
- **Post-Customer meeting**: Log Execution Notes and Update Risk Flags.

By Thursday afternoon, Northbridge had transformed marketing from a creative department into a revenue engine. The INSIGHT cadence delivered more than organized dashboards—it produced verified proof that specific audiences, messages, and channels drive incremental sales. Every campaign now generates field intelligence that sales could use in buyer meetings, complete with local performance data and verified growth metrics.

But precision marketing was only half the equation. The other half was ensuring that insights reached the people who could act on them most directly: the sales professionals sitting across from buyers who controlled shelf space, promotional calendars, and category growth investments. Marketing had proven what worked; now came the critical handoff that would determine whether those insights translated into expanded distribution, increased velocities, and stronger retailer partnerships.

The next phase would test whether Northbridge's data-driven approach could influence the conversations that mattered most, the ones that took place in buyers' offices, where they decided which brands earned more space, better positioning, and larger promotional investments. Success would be measured not in campaign metrics, but in the incremental shelf movement that only retailers could deliver.

Chapter Fourteen

Insight-Driven Sales Acceleration:

When Evidence-Based Selling Replaces Relationship Pitching

Sales execution transitions from relationship-dependent pitching to evidence-driven revenue planning, leveraging verified performance data, predictive account targeting, and measurable outcome tracking. This chapter provides the insight-driven sales framework that turns first-party customer signals into accelerated buyer decisions through systematic account prioritization and execution accountability.

We'll examine four sales levers that convert marketing insights into negotiation advantages, introduce territory allocation models that maximize resource efficiency, and provide objection-handling frameworks that ensure consistent sales velocity—the speed at which verified insights translate into expanded distribution and shelf space. You'll learn to establish measurement protocols that track expansion success—the verified growth in account value attributable to data-driven selling—and execution systems across the complete sales lifecycle.

By the end of this chapter, you'll have operational blueprints for implementing insight-driven sales processes, account prioritization frameworks that maintain focus while scaling territory coverage, and measurement systems that prove sales ROI while accelerating buyer confidence and partnership depth.

Northbridge: Data-Backed Selling

By the time Northbridge's marketing insights reach the sales team, they're already battle-tested in the market. Every chart, lift calculation, and cohort analysis from the INSIGHT framework becomes a sales asset, proof that campaigns engage consumers and drive products off shelves. This seamless handoff lets sales teams build on verified outcomes.

This approach comes to life in a Friday midday buyer meeting: regional account lead Carla Brooks opens with a live heatmap of stores showing promotions that have driven over 5% incremental lift. As the dashboard updates in real time, buyers lean forward. This isn't a sales pitch; it's a revenue report tailored to their locations.

When retailers can see exactly how a campaign performed in their stores, the conversation shifts from "Why should we?" to "How fast can we scale this?"

Key takeaway: Providing real-time, local performance evidence leads to accelerated buy-in, shifts buyer questions to focus on scaling, and results in actionable interest from retailers.

From Coverage to Conversion

Traditional sales teams aim for maximum coverage, often pushing the same pitch to every retailer. With insight-driven sales, reps prioritize accounts most likely to grow and use data-backed stories, leading to sharper targeting and better outcomes.

From Spray-and-Pray to Precision Selling:

Before diving into tactics, consider how first-party insights transform the traditional sales playbook. The following table contrasts Northbridge's legacy approach with its data-driven upgrade. As you review, identify any lingering 'Traditional Method' practices in your process and note the 'Insight-Driven Method' alternative to test next quarter.

Key takeaway: Select one specific data-driven practice to pilot and evaluate its impact next quarter.

Traditional Method	INSIGHT Driven Method
Call on every account equally	Rank accounts by expansion propensity & retention risk
Lead with national marketing talking points	Tailor pitch to local purchase data & consumer feedback
Negotiate on price alone	Negotiate on proven incremental lift and category growth
Ask for a small placement order as a favor to the Sales Representative	Show consumer purchase interest captured at local tastings to justify large displays, premium shelf positions, and feature placement
React to stock issues reported by stores	Proactively display compliance & stockouts via field data

Transition to Precision Selling

Four Sales Levers Powered by Insight

First-hand proof turns selling into planning. Instead of sharing the same message everywhere, representatives focus on four key areas: who they contact, what they offer, how they handle pricing, and where they spend their time in stores. Turn INSIGHT reports into actions for each account. Adjust one area at a time, compare the results, and repeat the process for what works.

1. **Account Prioritization** – Predictive scores identify retailers that show rising consumer demand and cross-category attachment potential.

2. **Pitch Relevance** – Receipt data reveals which flavor bundles resonate regionally, informing tailored assortment proposals.

3. **Negotiation Leverage** – Verified lift from QR-enabled promotions shifts conversations from cost to growth.

4. **Field Execution Clarity** – Photo-verified display compliance and stock-out alerts focus rep time where execution gaps threaten sales.

Turning Objections into Opportunities with First-Party Proof

Even with strong data, buyers may object. Now Northbridge's reps reply with specific, verifiable responses backed by first-party insights. Here's how common pushbacks are addressed:

1. **"Your product doesn't move fast enough in our stores."**

    - **Traditional response:** "Give us another chance with better placement."

    - **Data-driven response:** "Our receipt data shows 67% of purchasers in your region buy between 3-5 PM on weekdays. Your current shelf placement is in aisle 7, but our heatmap data indicates customers who engage with our brand typically shop aisles 2-4. Moving our product there could increase velocity by 30% based on similar store reconfigurations."

2. **"The category is declining."**

    - **Traditional response:** "Our brand is different."

    - **Data-driven response:** "You are correct about category decline, but our first-party data identifies a growing segment of 'wellness-conscious convenience seekers' who are three times more likely to buy natural products. Here is the demographic profile for your trade area, showing 2,400 high-intent consumers who have interacted with our wellness content and have yet to find us in your stores."

1. **"Your competitor offers better margins."**
   - **Traditional response:** "But our quality is superior."
   - **Data-driven response**: "Let's examine value beyond front margins. Our verified purchase data reveals customers have basket values 40% higher and shop 2.3 times more often than the category average. This lifetime value analysis for your format shows $47 greater profit per customer annually, despite the lower unit margin."
2. **"We don't have space for new SKUs."**
   - **Traditional response:** "Can you make room?
   - **Data-driven response:** "Our sales velocity analysis of your current set shows three SKUs averaging less than 1 turn per week. Our test markets with similar demographics achieve an average of 4.2 turns per week. We're proposing a one-for-one swap that could increase your revenue per linear foot by 22%."

Northbridge Case: Regional Convenience Chain Upsell

This is the playbook in motion: by using sales and purchase data to find where demand is highest, the team created a focused offer and supported it with store-level facts. The result wasn't a longer pitch, it was a faster "yes" and measurable growth on the shelf.

- **Insight:** Consumers in the chain's urban stores redeemed QR rebates at twice the national average and had higher repeat purchase intervals.
- **Action:** Sales proposed adding two companion SKUs and implementing a quarterly in-store sampling program with data capture.
- **Outcome:** The chain agreed to expand shelf space by 15%. Three

months after the rollout, category sales at test stores increased by 11% according to POS data. Key takeaway: The data-driven expansion of shelf space, as evidenced by a 15% increase and an 11% sales lift, demonstrates that performance-based programs can measurably improve sales outcomes.

Strategic Territory Allocation Using First-Party Signals

Rather than dividing territories solely by geography, Northbridge now allocates sales resources based on opportunity density, a combination of current performance and predictive growth potential.

Opportunity Scoring Framework:

Each territory receives a composite score based on:

- Current Performance (30%): YoY growth, market share, distribution gaps
- Consumer Signals (40%): First-party engagement density, purchase intent scores, unfulfilled demand indicators
- Competitive Dynamics (20%): Share of voice, promotional intensity, switching signals
- Execution Quality (10%): Compliance rates, stock-out frequency, merchandising effectiveness

Resource Allocation Model:

- **Tier 1:** "Growth Engines" (Top 20% of opportunity scores): Weekly visits, dedicated account teams, first access to innovation

- **Tier 2:** "Steady Builders" (Middle 50%): Bi-weekly visits, standard support, promotional focus
- **Tier 3:** "Maintain & Monitor" (Bottom 30%): Monthly visits, automated reordering, efficiency focus

Dynamic Rebalancing:

Every quarter, Northbridge's sales operations team reviews territory scores. When a Tier 3 account shows sustained engagement growth (like a 20% increase in local QR scans), it triggers automatic elevation to Tier 2, with corresponding resource reallocation. This system led to a 15% improvement in sales productivity by focusing effort where data predicted the highest returns.

Example: The Phoenix territory jumped from Tier 3 to Tier 1 after first-party data revealed a 150% spike in "where to buy" searches and unfulfilled online orders. The increased sales focus resulted in securing three new major accounts within 60 days.

Sidebar – Real-world example

Promolytics arms sales teams with store-level dashboards that directly connect promotional engagement to verified sales. Today, the platform captures data from multiple channels, **in-store tastings, events, shelf talkers and posters at retail, on-premise menus, TV commercials, cookies, social media, print ads, and emails,** updating the unique customer profile with each interaction while preserving source attribution.

Sales professionals arrive at buyer meetings with dashboards, not anecdotes. For example: "Promotional tastings in your zip code attracted 500+ potential customers for our new line extension." This is proof, backed by transaction data.

The immediate impact:

- **Negotiation Leverage** – Buyers are far more willing to expand SKUs or shelf space when they see measurable revenue upside. Instead of just asking for *three bottles to trial a product*, a representative can point to local tasting and promotional results. This allows them to confidently request **floor case stacks and a #1 display position**, knowing the data supports the potential sell-through.

- **Tailored Recommendations** – Sales teams can propose assortments or bundles based on verified consumer preference data, rather than relying on guesswork.

- **Attribution Confidence** – Tie consumer engagement directly to sales outcomes, shifting conversations from cost to growth.

- **Expanded Data Capture** – Incorporating tasting events, in-store signage, and on-premise activations into the same unified view.

- **AI-Powered Predictive Insights** – Anticipate which SKUs, promotions, and timing will yield the highest sell-through for a specific store footprint.

- **Execution Accountability** – Reps can show photo-verified compliance and tie it directly to performance changes, keeping the conversation results-focused. Key takeaway: Using performance-linked accountability keeps teams focused on actions that drive impact.

What's coming next:

- **Cross-Department Alignment** – Marketing, sales, and product teams work from the same performance data, ensuring that one team's delay or success is immediately visible to all.

- **Revenue-Linked Accountability** – Every milestone is linked to downstream metrics, such as incremental lift, match rate, or churn reduction, ensuring discussions remain grounded in results.
- **Real-Time Corrections** – If a metric dips below target, Promolytics will automatically flag the owner and relevant teams, prompting rapid intervention, often within the same day. *Example:* During a recent beverage category pilot, the sales team saw a dip in engagement in Week 6. The platform flagged the drop, linked it to a specific in-store display compliance issue, and sent a task to field reps. The issue was resolved within 72 hours, preventing a predicted 4% revenue loss for the quarter.
- **AI-Enhanced Execution** – The next iteration will not only detect issues but also suggest corrective actions proven to deliver the best turnaround in similar situations, turning the 90-day plan into a self-optimizing execution engine.

The Connected Rep: Digital Tools That Turn Insights into Field Actions

Northbridge equipped every sales rep with a mobile-first toolkit that brings first-party insights directly to the point of sale:

The Sales Intelligence App:

Pre-Visit Intelligence:

- Account-specific dashboards refresh the morning of the scheduled meetings
- Competitive activity alerts based on consumer switching signals

- Suggested talk tracks based on recent local campaign performance
- Inventory alerts flagging potential stock-outs

In-Store Capabilities:

- Real-time shelf photo capture with AI-powered compliance scoring
- Competitor price scanning with instant margin impact calculations
- Location-based consumer heat data ("47 high-intent shoppers within 1 mile")
- One-click ordering for identified gaps

Post-Visit Actions:

- Automated visit reports with photo documentation
- Follow-up triggers based on commitment tracking
- Performance updates are sent to buyers within 24 hours
- Next-visit planning based on execution gaps

Integration Impact:

Reps using the full digital toolkit achieved:

- 32% reduction in administrative time
- 28% increase in average order value
- 45% improvement in promotion compliance
- 18% faster sales cycle completion

Training Investment: Northbridge required every representative to complete a certification program, with refresher sessions provided as new features were introduced. The initial productivity dip lasted two weeks before gains materialized.

Executive Summary - Sales Technology ROI: Digital sales tools aren't productivity accessories—they're competitive necessities. Reps using integrated insight platforms achieve 32% time savings, 28% higher order values, and 45% better execution compliance. The key insight: technology amplifies preparation and proof, turning every sales conversation into an evidence-based revenue discussion rather than relationship-dependent persuasion.

Key Metrics for Sales Effectiveness

Putting numbers behind the pitch: Sales credibility depends on measurable outcomes. The scoreboard below uses four KPIs that every account manager reports to leadership: revenue growth, sales cycle speed, execution quality, and impact of insights. Each metric directly shows progress in these areas, proving the insight engine's contribution.

Metric	Why It Replaces Old Metrics	Northbridge Target
Expansion success rate	Focus on growth within existing accounts, not just new logos	≥ 20% of Tier A accounts expand annually
Average days to close	Reflects efficiency driven by better prioritization	Cut the cycle from 60 to 45 days
Verified incremental lift per promotion	Shifts negotiation from discounts to value creation	≥ 6% lift at participating retailers
Execution gap resolution time	Ensures field issues are addressed quickly	Resolve 90% within 72 hours

Key Metrics

Incentives That Drive Insight Adoption

Of course, for such changes to succeed, teams need to be aligned. Northbridge responded by integrating insight use directly into incentives:

Revised Compensation Structure:

1. Base (50%): Unchanged

2. Traditional Metrics (25%): Revenue, new distribution

3. Insight Metrics (25%):

   - Predictive account score improvements
   - Data tool utilization rates
   - Compliance photo submissions
   - Lift vs. holdout performance

This shift initially met resistance, but within two quarters, reps leveraging insights consistently out-earned those using traditional methods by an average of 18%, creating natural momentum for adoption.

Action Steps Implemented

- Defined **per-account OEC** (e.g., incremental units vs. holdout, payback) and stored it in the CRM; every pitch aligns to that OEC.
- Auto-generated **Buyer Lift Cards** (1-pager) for each meeting: baseline, design, holdout, cost per incremental unit, next ask.
- Set Execution Gap → Action SLAs: stock-out alert → send buyer email within 2 hours; non-compliant display → create ticket for merch team within 24 hours.
- Recalculated the **territory opportunity score monthly**; auto-tiered accounts (promote/demote) with predefined resource plays.
- Added **predictive expansion score** to the Sales CRM, updated periodically.
- Embedded store-level **lift cards** in pitch decks.

- Set an **SLA** for field reps to upload shelf photos within 24 hours of the visit.
- Published an **Objection → Proof library** (talk tracks, charts, and SKUs to swap) inside the sales app; updated monthly for new products or quarterly for established SKUs.
- Created a **save-play** for accounts with declining sell-through, signaled by engagement drop.
- Created a **Field Experiment Registry** (unique IDs, controls, results) to avoid re-running the same test.
- Launched a **partner portal** for top accounts (store-level lift, demand heatmaps, promo calendar).
- Scheduled **monthly alignment call** between sales and marketing to share promotion performance insights.

Cross-Functional Implementation: Sales Intelligence Accountability Matrix

Insight-driven sales requires coordination across every function that influences buyer relationships and account growth. Unlike traditional sales approaches that rely on individual rep relationships, data-driven selling embeds systematic intelligence gathering and sharing throughout the revenue process. Successful account expansion requires coordinated intelligence that buyers can trust. When opportunity signals emerge, the entire organization mobilizes to support the sales conversation with verified proof rather than unsubstantiated claims. The following accountability matrix ensures buyer meetings feature consistent, credible evidence rather than isolated anecdotes.

- **Sales Operations:** Owns account scoring models, territory optimization, and performance analytics. Maintains predictive algorithms and resource allocation protocols across all accounts.

- **Marketing:** Provides campaign performance proof, local market insights, and promotional effectiveness data. Creates account-specific materials and validates the quality of the marketing-to-sales handoff.
- **Field Execution:** Manages compliance monitoring, inventory tracking, and resolution of execution gaps. Provides real-time field intelligence and photo verification systems.
- **Category Management:** Analyzes assortment performance, shelf optimization opportunities, and competitive dynamics. Supports data-driven product placement and merchandising recommendations.
- **Customer Success:** Monitors account satisfaction, processes buyer feedback, and identifies relationship risk factors that could impact expansion opportunities.

From Annual Handshakes to Dynamic Partnership Planning

Traditional joint business planning meant annual meetings with static projections. Northbridge now runs quarterly business reviews powered by shared first-party insights:

The Collaborative Dashboard:

Northbridge provides key accounts with customized portals showing:

1. Real-time category performance by store
2. Consumer demand signals for their specific trade areas
3. Promotional performance with ROI calculations

4. Predictive alerts for trending opportunities

Quarterly Planning Rhythm:

Month 1: Data Gathering

1. Aggregate the previous quarter's performance
2. Analyze consumer behavior shifts
3. Identify execution gaps and opportunities

Month 2: Strategic Alignment

1. Joint review of insights with category buyers
2. Collaborative target setting based on predictive models
3. Resource allocation agreements

Month 3: Execution Sprint

1. Launch aligned initiatives
2. Daily performance monitoring
3. Real-time optimization based on early signals

Success Metrics from Joint Planning:

Accounts participating in data-driven joint planning showed:

- 24% higher annual growth vs. non-participating accounts
- 60% faster new item authorization
- 35% reduction in out-of-stocks

- 2.3x promotion effectiveness

Case Example: A regional chain used Northbridge's consumer insights to identify a gap in afternoon snacking occasions. Together, they created a "3 PM Pick-Me-Up" endcap program that drove 18% category growth in participating stores. Key takeaway: Acting on local demand signals leads to targeted programs and measurable growth.

Executive Reflection – Growth Focus:

Which account currently consumes the most sales effort with the least verified incremental growth, and how can insights reallocate resources to higher-potential opportunities?

Chapter 14 Checklist: Insight Driven Selling

- **Lock an OEC** for top accounts and baseline it before submitting proposals; include a holdout design in every deck.
- Rank all accounts by **expansion propensity** and **retention risk.**
- Equip the top 15% of accounts with **data-backed pitch decks** showing verified lift.
- **Conduct at least one controlled in-store test per quarter** in each Tier 1 account before rolling out the chain-wide rollout.
- Implement **shelf photo verification** and tie compliance to an incentive.
- **Buyer recap rule:** send commitments/next steps after every meeting; track follow-through rate.
- Launch **save-play triggers** for accounts with declining engagement signals.

- **Monthly win/loss review** with root cause (assortment, execution, price, competitor) and a countermeasure logged.
- **Objection library refresh** with the latest proof points and talk tracks.
- Review **execution gap resolution times** and establish improvement targets to optimize performance.

By the end of the quarter, Northbridge had transformed both marketing precision and sales effectiveness into competitive advantages. However, their success revealed a new opportunity: with the ability to identify exactly which customers wanted specific products, and sales teams equipped to secure optimal shelf placement, the next logical step was to ensure they were offering the right products in the first place.

The data was clear—some product variants consistently outperformed expectations while others struggled despite perfect execution. Customer signals revealed unmet needs, and regional preferences showed significant variation that standard assortments couldn't address. The question wasn't whether to expand their product strategy, but how quickly they could align their innovation pipeline with the customer insights they were now capturing systematically.

The next chapter reveals how first-party data transforms product development from intuition-based launches into market-validated innovations that customers actively request before they even reach the shelves.

Chapter Fifteen

Data-Driven Product Wins:

How First-Party Insights Guide Innovation

Product development transitions from intuition-based feature creation to evidence-driven innovation through the detection of behavioral signals, rapid validation cycles, and the tracking of measurable outcomes. This chapter presents an insight-led product framework that converts first-party consumer signals into profitable launches through systematic concept validation and performance measurement.

We'll examine four product levers that convert consumer behavior into development priorities, introduce validation models that accelerate concept-to-market cycles, and provide decision frameworks that ensure consistent innovation velocity—the speed at which validated consumer insights translate into scalable product offerings. You'll learn to establish measurement protocols that track repeat velocity—the verified adoption rate of new concepts among target segments—and iteration systems across the complete development lifecycle.

By the end of this chapter, you'll have operational blueprints for implementing insight-driven product processes, concept validation frameworks that maintain consumer focus while scaling innovation capacity, and measurement systems that prove product ROI while accelerating market-validated development cycles.

Northbridge: From Signal to SKU

Monday in Northbridge's product studio looks different now. Where trend decks and competitor teardowns once dominated wall space, live dashboards display first-party behavioral signals from QR surveys, receipt uploads, and partner sampling events. Prototypes of new snack sizes and beverage flavors still line the shelves, but each concept now carries a data story—not just aesthetic appeal.

"We're done building products people might want," says Troy Olson, the head of product innovation, wheeling in a display of real-time consumer feedback. "We build what behavioral signals prove they'll buy again."

The shift from intuition-driven to evidence-based product development doesn't happen overnight, but the competitive advantage compounds quickly. Teams that master first-party insight generation can spot demand inflections 3-6 months before they appear in industry reports, and validate concepts in weeks rather than quarters.

From Feature Factory to Insight-Led Innovation

Traditional product development operates on delayed feedback loops, relying on annual trend reports, quarterly focus groups, and post-launch sales data that arrives too late to course-correct. By then, shelf space is committed, inventory is produced, and marketing budgets are allocated around unproven assumptions.

Insight-driven development flips this equation. Consumer behavior becomes the primary input signal, verified purchases validate concept appeal, and actionable feedback shapes iteration cycles in real-time. The roadmap evolves from a collection of hoped-for hits into a portfolio of performance-verified bets.

Consider how your current process measures up:

Traditional Approach:

- Annual trend reports drive concept ideation
- Focus groups provide directional feedback

- Launch-and-learn with national rollouts
- Success is measured by initial sales velocity
- Post-mortem analysis after market performance

Insight-Driven Approach:

- Behavioral signals reveal unmet needs continuously
- Receipt-verified pilots confirm purchase intent
- Iterative validation before scale commitment
- Success is measured by repeat velocity and margin expansion
- Real-time optimization throughout launch cycles

The transformation requires discipline: every concept must earn its place through consumer validation rather than internal enthusiasm or competitive mimicry.

What This Isn't

This approach doesn't eliminate intuition or creative leaps; it validates them faster. Product leaders still need vision to spot emerging opportunities and creativity to design compelling solutions. The difference is that hunches get tested within weeks rather than months, and resources flow toward concepts that demonstrate real consumer traction rather than internal enthusiasm.

This isn't about analysis paralysis or waiting for perfect data before acting. The goal is rapid validation cycles that either confirm your thesis or redirect resources toward better opportunities. Great product leaders combine pattern recognition with systematic testing, using behavioral signals to separate genuine insights from wishful thinking.

Four Product Levers Powered by Insight

Converting consumer signals into profitable products requires more than collecting feedback; it demands systematic frameworks that transform behavioral data into actionable development priorities. Most product teams drown in consumer input without clear mechanisms to separate genuine demand signals from temporary preferences or vocal minorities.

The four levers below create a repeatable process for translating first-party insights into product decisions. Each lever addresses a specific stage of the validation journey: detecting early demand, prioritizing friction points that matter, measuring outcomes that predict success, and testing concepts before committing resources. Together, they form an evidence engine that builds exponential insights with every iteration.

1. **Demand Signal Detection** – Consumer preference polling and cross-category exploration reveal early interest in adjacent flavors, formats, or use cases before they appear in broad market data. QR surveys at the point of purchase capture immediate reactions and articulate unmet needs. Partnering with sampling events in relevant contexts, such as commuter hubs for portable formats and fitness centers for performance variants, amplifies signal strength and validates contextual demand. Retail partners often provide access to high-intent consumers in natural usage environments, thereby accelerating signal quality compared to internal-only testing.

2. **Friction Prioritization** – Open-text feedback linked to specific drop-off points in the purchase journey pinpoints usability, packaging, or experience issues that suppress repeat adoption. Rather than generic satisfaction scores, friction analysis identifies the precise barriers preventing customers from becoming advocates. Partner retail locations offer unique insights into real-world friction points, such as shelf accessibility issues, checkout flow problems, or packaging challenges that only arise in actual store environments. This enables surgical improvements

rather than broad redesigns.

3. **Outcome Metrics That Matter** – Time to repeat purchase and adoption depth replace feature count and launch velocity as primary indicators of success. Customer lifetime value expansion, category penetration rates, and margin contribution per acquisition provide clearer ROI signals than traditional volume metrics alone. Partner channel data provides comparative baselines across different retail environments, helping validate whether performance improvements reflect genuine product appeal or location-specific factors.

4. **Iterative Validation Cycles** – Micro-tests using limited store pilots, partner channel distribution, and receipt validation confirm demand patterns before committing to national rollout infrastructure. Partner networks can reduce validation timelines by 30-40% through parallel testing across complementary channels, including convenience stores, gyms, and transit locations, providing diverse consumer context without requiring internal infrastructure investment. This approach reduces launch risk while preserving upside potential by rapidly scaling validated concepts.

Executive Summary - Innovation Acceleration: The four-lever framework transforms product development from lengthy research cycles into rapid validation engines. Demand signal detection identifies opportunities months ahead of competitors, friction prioritization focuses resources on problems that matter, outcome metrics prevent feature bloat, and iterative validation reduces launch risk while preserving upside potential. Companies using this approach typically cut concept-to-market time by 40-60% while improving success rates.

Northbridge Case: Slim Can Launch Decision

Theory sounds compelling in conference rooms. Proof emerges in checkout lines and in moments of consumption. When Northbridge's product team first heard requests for smaller packaging formats, the feedback felt anecdotal, coming from a few vocal customers who wanted something different. Traditional market research would have taken months to validate demand and cost several thousand dollars to execute properly.

Instead, the team deployed their four-lever framework to transform scattered consumer comments into a validated product launch within 12 weeks. The slim can case demonstrates how first-party signals, when properly captured and analyzed, can reveal profitable opportunities that are often hidden in plain sight. More importantly, it shows how rapid validation cycles enable teams to act on consumer insights while competitors are still commissioning research studies.

Here's how demonstrated demand signals guided every decision from concept to regional rollout:

- **The Signal:** QR surveys deployed at urban convenience stores revealed that 28% of commuter shoppers prioritized portability over volume, specifically mentioning the difficulty of fitting standard 12-ounce cans into laptop bags and gym totes. Partner sampling events at transit hubs amplified this signal, with 34% of participants requesting smaller format options.

- **The Friction:** Sentiment analysis of open-text feedback highlighted portability frustrations with existing packaging, particularly among the 25-40 demographic that represented the brand's highest lifetime value segment.

- **The Test:** Product teams piloted an 8-ounce slim can format in 50 stores across three urban markets, supplemented by targeted sampling through two transit-focused retail partners. Receipt validation tracked actual purchase behavior rather than stated intent, while post-purchase surveys captured satisfaction and repeat likelihood scores.

- **The Result:** Verified incremental lift of 9% in pilot locations, with 23% higher repeat purchase rates among commuters compared to standard format buyers. Regional expansion decision made within 12 weeks of initial pilot launch, supported by margin analysis showing 15% higher profitability per unit despite reduced volume.

- **The Multiplier Effect:** Partner collaboration accelerated validation by nearly 40%, with retail insights providing complementary data to internal channels, thereby enhancing the overall effectiveness of the process. This hybrid approach reduced the total validation timeline while increasing confidence in scaling decisions.

When Signals Say Stop: The Premium Mixer Misstep

Not every consumer request translates to a market opportunity. When Northbridge's QR surveys revealed that 19% of craft cocktail enthusiasts requested a premium mixer line, "something sophisticated for home bartending," the initial signal seemed compelling. Upscale restaurant partnerships reinforced the concept's potential, with bartenders expressing interest in co-branded specialty mixers.

But the four-lever framework revealed critical gaps. **Demand Signal Detection** showed interest was geographically concentrated in just three metropolitan areas, representing less than 8% of Northbridge's total market footprint. **Friction Prioritization** uncovered price sensitivity concerns during blind sampling, with 73% of participants unwilling to pay the premium required to achieve target margins on small-batch production.

Most telling, the **Iterative Validation** pilot in 15 upscale grocery locations yielded initial trial rates of only 2.3%, with virtually no repeat purchases after 45 days. Post-purchase interviews revealed customers viewed premium mixers as "special occasion" purchases rather than regular consumption items, fundamentally misaligning with Northbridge's volume-based business model.

The team halted the project after ten weeks, disregarding the enthusiasm of cocktail aficionados, and redirected resources to the slim can opportunity, which demonstrated consistent repeat purchases across demographics.

The lesson: Strong signals need real purchase behavior, repeat potential, and fit with your business model. The framework protects you from costly missteps as much as it finds breakthrough ideas.

The Metrics That Stop Feature Bloat

Shipping features feel productive. Proving they create lasting value requires discipline. Northbridge focuses on four core metrics to separate genuine innovation from internal enthusiasm:

Repeat Velocity: Time from initial purchase to second purchase, segmented by acquisition channel and customer profile. Products that drive faster repeat cycles indicate stronger product-market fit.

Adoption Depth: Percentage of trial customers who become regular purchasers within 90 days. This metric reveals whether initial interest translates to sustained demand.

Margin Expansion: Contribution per customer over 12 months compared to baseline products. Innovation should improve unit economics, not just volume.

Friction Resolution: Reduction in drop-off rates at specific journey stages after product improvements. This tracks whether enhancements solve real customer problems rather than internal assumptions.

These metrics create natural stopping points for concepts that generate internal excitement but fail to drive measurable changes in customer behavior.

Building Your Evidence Engine

Immediate Actions:

- Add "unmet needs" capture fields to existing customer touchpoints (QR surveys, receipt uploads, support interactions)
- Establish rolling 8-12 week pilot validation cycles for new concepts
- Define minimum lift thresholds and repeat velocity requirements for scaling decisions
- Integrate product validation metrics into regular leadership reviews
- Archive low-adoption variants to focus resources on high-potential opportunities

Sidebar – Real-world example

Promolytics enables brands to test, measure, and refine product concepts before scaling them up. Today, the platform captures data from QR codes on packaging, digital ads, in-store tastings, partner-led sampling events, receipt uploads, and interactive links in SMS or email campaigns, combining each interaction into the unified consumer profiles while preserving source attribution. The team can use this database to segment and analyze the data to provide ranked concept viability, segment-level purchase intent, price sensitivity and offer response, region heatmaps, and clear go, iterate, or scale recommendations tied to verified outcomes.

For product teams, this means you can track not just *if* a SKU is selling, but *where, how, and why*. Takeaway: Real-time insights enable you to quickly adjust production, marketing, and rollout for maximum effectiveness.

Where it's going: Future AI and ML integrations could predict the most promising product concepts *before* they hit production, modeling potential demand based on historical sales performance, demographic match rates, and cross-category adoption curves. This would enable brands to green-light winners and avoid costly misfires.

Action Steps Implemented

- Partnered with three retail locations to provide parallel validation environments and reduce testing timelines
- Added a "future ideas" feedback field to QR surveys capturing unmet needs.
- Established a rolling twelve-week pilot schedule for concept validation.
- Defined a minimum lift threshold and repeat velocity KPI for scaling decisions.
- Integrated product feedback metrics into the INSIGHT Wednesday Interpret session.
- Archived low-adoption legacy variants to free shelf space and focus on high-potential formats.

Cross-Functional Implementation: Product Innovation Accountability Matrix

Evidence-driven product development requires coordination across every function that influences innovation success and market validation. Unlike traditional development approaches that operate in R&D silos, insight-led innovation embeds consumer validation throughout the entire development process. Product concepts that survive market contact look fundamentally different from those that emerge from internal brainstorming. When consumer behavior reveals unmet needs, the organization transforms insights into testable prototypes through disciplined validation rather than speculative feature development. This framework ensures that product development follows consumer-proven demand rather than internally generated enthusiasm.

Product Management: Owns concept prioritization, validation, design, and outcome measurement. Maintains behavioral signal analysis and concept performance tracking across all development initiatives.

Marketing: Provides consumer insight validation, pilot campaign execution, and adoption measurement. Creates validation environments and manages consumer feedback collection systems.

Operations: Manages pilot production, retail partner coordination, and supply chain flexibility. Enables rapid concept testing and scaling protocols.

Data Analytics: Develops predictive models for concept success, maintains validation measurement integrity, and provides performance attribution across innovation cycles.

Sales: Provides retail partners with access, field validation support, and integration of buyer feedback. Manages pilot placement and execution monitoring.

Executive Action:

Identify your highest-conviction upcoming launch and apply the four-lever framework: What behavioral signals support demand? Which friction points would prevent repeat purchases? What outcome metrics will determine success? Design a 30-day pilot with receipt validation that tests these assumptions.

The goal isn't to eliminate all uncertainty. Instead, it's to fail faster on weak concepts and scale aggressively when consumer behavior validates your thesis. This approach accelerates learning, enables faster resource reallocation, and boosts your team's confidence in decision-making. Each validation cycle builds institutional knowledge about what drives repeat purchases in your category. This creates compounding advantages over competitors who rely on delayed market feedback.

Great products emerge when consumer behavior becomes your primary compass. First-party data then enables rapid iteration toward product-market fit. Companies that master this approach don't just launch more successful products; they also achieve greater market share. They build innovation engines that get stronger with each cycle.

Chapter 15 Checklist: Insight-Driven Product Development

- Establish partner validation channels for accelerated testing cycles.
- Create failure criteria and exit thresholds for each pilot concept.
- Launch a micro user experience targeting unmet needs in key segments.
- Define outcome metrics (repeat velocity, adoption depth) for all new concepts.
- Select one upcoming idea and design a limited receipt validated pilot.

- Incorporate friction driver analysis into the monthly product review.
- Retire or redesign one low-adoption feature to reclaim resources.

By the end of the slim can pilot, Northbridge didn't just launch a successful product—they established a repeatable blueprint for turning raw signals into profitable bets. Sustaining this momentum, though, requires continued discipline beyond the pilot phase.

Leadership must anticipate market inflection points before trends become visible. Evelyn recognizes that robust data infrastructure powers both near-term gains and detects early shifts or threats. Success depends on leaders who act decisively on emerging signals, ahead of competitors.

As the product team celebrated their regional rollout success, the executive team gathered for a different conversation: not about the next flavor, but about anticipating the company's next inflection point.

Chapter Sixteen

Signal Intelligence:

How Leaders Capture Tomorrow's Opportunities

Strategic leadership evolves from reactive management to predictive market positioning through the interpretation of early signals, scenario modeling, and accelerated decision-making cycles. This chapter presents an insight-driven leadership framework that converts first-party consumer and partner signals into competitive advantages through systematic opportunity identification and rapid organizational response.

We'll examine four leadership levers that convert weak market signals into strategic advantages, introduce decision acceleration models that compress traditional planning cycles, and provide alignment frameworks that ensure consistent leadership velocity—the speed at which emerging opportunities translate into coordinated organizational action. You'll learn to establish measurement protocols that track signal-to-strategy lag—the verified time between opportunity detection and market response—and competitive positioning systems across the complete leadership lifecycle.

By the end of this chapter, you'll have operational blueprints for implementing predictive leadership processes, signal interpretation frameworks that maintain competitive awareness while scaling decision speed, and measurement systems that prove strategic ROI while accelerating market timing and opportunity capture rates.

Northbridge: From Signal to Strategy

Monday afternoon in the executive conference room. The quarterly review has just concluded with largely positive results, as revenue targets were met, campaign performance was solid, and the slim can launch exceeded projections. As department heads begin packing laptops and celebrating wins, Evelyn remains focused on her dashboard, scrolling through real-time consumer sentiment data.

"This is exactly when we're most vulnerable," she says, addressing the room as conversations quiet. "Success creates blind spots. While we're congratulating ourselves on last quarter's performance, early signals are already forming that will determine whether we're leading or following in six months."

She pulls up a split screen showing their current metrics alongside emerging consumer behavior patterns that won't appear in competitor reports for another 60 to 90 days. "Traditional leadership celebrates yesterday's wins. Insight-driven leadership uses today's momentum to capture tomorrow's opportunities."

This distinction, between managing current success and anticipating future inflection points, defines the difference between reactive and predictive leadership in an insight-driven organization.

Four Leadership Levers Powered by Insight

1. **Early Signal Interpretation** – Deploy automated alerts that synthesize consumer sentiment shifts, regional demand fluctuations, and partner channel performance changes into executive-level intelligence. This requires integrating social listening, sales velocity tracking, and partner feedback loops into unified dashboards that highlight pattern changes before they impact quarterly results.

2. **Scenario Modeling** – Build decision trees that evaluate multiple "what if" cases using historical data and predictive analytics. Model supply chain constraints, regulatory changes, competitive responses, and market shifts to prepare response strategies before events unfold. This trans-

forms reactive crisis management into proactive opportunity positioning.

3. **Cross-Functional Alignment** – Establish shared data protocols to ensure marketing, sales, product, and finance operate from identical forward-looking insights, rather than conflicting departmental reports. Weekly alignment sessions review emerging signals and coordinate response strategies, preventing siloed reactions that waste resources and confuse market execution.

4. **Opportunity Acceleration** – Create fast-track decision protocols that compress standard approval cycles when high-confidence signals indicate time-sensitive opportunities. Pre-approved budget allocations and decision-making authorities enable teams to transition from signal detection to market action within weeks, rather than quarters.

These four levers can represent a fundamental shift in how executives operate. The contrast becomes clear when you examine traditional versus insight-driven approaches across key leadership decisions:"

Traditional Leadership	Insight-Driven Leadership
React to market shifts after they're visible in sales reports	Monitor leading indicators in consumer behavior and competitor response
Base strategic bets on industry trend reports	Layer partner data, pilot results, and predictive modeling for decision-making
Wait for annual planning cycles to adjust	Hold rolling strategy checkpoints tied to real-time KPIs
Measure success on past performance	Measure readiness for future opportunities

Leadership Transformation Summary

Northbridge Case: The Wellness Pivot That Saved Q4

The Context: September 2024. Northbridge's core beverage sales had plateaued for two consecutive quarters while production costs climbed 8% due to supply chain pressures. Traditional leadership would have focused on cost-cutting and promotional pricing to defend market share. Instead, Evelyn's team was tracking an emerging signal that would reshape their entire Q4 strategy.

The Signal: Meanwhile, social sentiment analysis revealed a 340% increase in wellness-related keywords associated with beverage searches over a 90-day period, with a concentration in suburban markets among individuals aged 28-45. Partner tasting events revealed that 67% of participants inquired about "functional benefits" rather than just flavor preferences. Most telling, QR survey data indicated that 41% of their highest-value customers were actively seeking "clean energy" alternatives to traditional caffeinated beverages.

The Leadership Challenge: Three of four department heads viewed this as a temporary health trend, similar to previous fads that had faded within months. The traditional response would have been to commission a six-month market study and plan a 2025 product line extension. Finance projected that any new product launch would require 18 months to break even.

The Accelerated Decision Process: In response, leadership used their four-lever framework to compress what would typically be a 12-month process into six weeks:

- **Early Signal Interpretation:** Real-time dashboard integration revealed the wellness trend was accelerating, not stabilizing, with cross-category validation from sports nutrition and functional food sectors

- **Scenario Modeling:** Predictive analytics showed three potential outcomes: ignore the signal (15% revenue decline risk), slow response (maintain status quo), or fast response (18-32% upside potential)

- **Cross-Functional Alignment**: Marketing, product, and sales aligned on shared data showing this wasn't a niche trend but a category shift affecting their core demographic

- **Opportunity Acceleration:** Leadership activated pre-approved innovation budgets and fast-tracked supplier relationships to pilot test within 30 days

The Execution: Building on this accelerated alignment, Northbridge chose not to develop entirely new products. Instead, they reformulated two existing flavors with natural caffeine sources and added vitamin B12, marketing them as "Clean Focus" variants. Limited pilots launched in 75 suburban locations across four markets, with receipt validation tracking actual purchase behavior versus stated intent.

The Results:

- **Financial Impact:** 23% revenue lift in pilot locations within 60 days, with 41% higher margin per unit due to premium positioning

- **Market Timing:** Beat three major competitors to market by 4-6 months, capturing first-mover advantage during peak wellness season

- **Strategic Value:** Established wellness credibility that positioned Northbridge for broader category expansion in 2025

- **Risk Management:** Total investment of $340K versus traditional new product development costs of $2.1M+

The Competitive Advantage: By December, when competitors finally launched their wellness lines, Northbridge had already secured premium shelf placement, established consumer trial, and built repeat purchase momentum. What began as scattered social sentiment had become a $4.2 million revenue boost, turning a flat Q4 into their strongest quarter on record.

The Leadership Lesson: Ultimately, the difference wasn't better data; competitors had access to similar market intelligence. The advantage came from leadership's ability to interpret weak signals, model scenarios rapidly, align functions around shared insights, and compress decision cycles from quarters to weeks. While competitors debated whether wellness was sustainable, Northbridge was already scaling its response.

This case shows that insight-driven leadership is a decisive differentiator. Those who spot signals, interpret them quickly, and rally teams to act transform fleeting trends into lasting advantage. In a market where speed matters, proactive action, not perfect data, defines the winners.

Sidebar – Real-world example

Promolytics can't yet *predict* the future, but it gives leadership the tools to spot it forming. The platform aggregates campaign, sales, and product test data into a unified executive dashboard, enabling leaders to:

- **Spot leading indicators** — such as spikes in QR scans, flavor-specific survey interest, or partner event engagement by ZIP code, weeks before they show up in sales numbers.

- **Align cross-functional teams** — so marketing, sales, and product are reacting to the same forward-looking insights, not conflicting lagging reports.

- **Move from proof to prediction** — Today, Promolytics optimizes live campaigns and validates concepts. In the future, AI and ML integrations will model "most likely" outcomes, helping leadership simulate scenarios before committing budget.

This shift transforms executive decision-making from reactive post-mortems to proactive market shaping, from asking "What happened?" to answering "What will happen next if we act now?"

When Leadership Signals Mislead: The Voice Commerce False Start

Leadership signals can mislead. In late 2023, Northbridge's social listening identified a 280% surge in voice-activated shopping discussions, particularly regarding beverage purchases made via smart speakers. Industry reports forecasted that voice commerce would reach 15% of CPG sales within 18 months. Partner retailers confirmed major investments in voice-enabled ordering systems.

Leadership fast-tracked a voice commerce strategy, allocating $450K toward optimizing product listings for voice search and developing audio brand recognition campaigns. The predictive models suggested a significant first-mover advantage in an emerging channel that competitors were ignoring.

The Reality Check: After six months of investment, voice-driven beverage purchases accounted for less than 0.3% of total sales. Consumer behavior research revealed that while people discussed voice shopping enthusiastically, actual adoption stalled due to concerns about product substitutions and the verification of impulse purchases. The technology infrastructure was advancing faster than consumer habits were forming.

The Learning: Strong technological signals and industry enthusiasm don't automatically translate to shifts in consumer behavior. The framework worked correctly—it enabled rapid testing and faster failure recognition. Rather than doubling down on a multi-million-dollar commitment, Northbridge redirected resources toward proven channels within one quarter, thereby limiting total losses to the initial pilot investment. Total investment was limited to $450K rather than the originally planned $2.3M, demonstrating the framework's ability to fail fast and redirect resources.

This experience reinforced the importance of behavior-based validation over following technology trends, even when early indicators appear compelling and industry consensus supports the opportunity.

The Metrics That Enable Market Leadership

Leadership requires different metrics than tactical execution. Operational teams track conversion and campaign results. Executives, however, require measures of competitive positioning and future readiness. The four metrics below distinguish leaders who shape markets from those who merely react to them.

Signal-to-Strategy Lag: Days between detecting a significant market signal and implementing a coordinated organizational response. Northbridge targets under 21 days for high-confidence opportunities, compared to industry averages of 90 to 120 days. This metric indicates whether your decision-making infrastructure can keep pace with the rapid evolution of the market. Calculate by tracking timestamps from the initial alert to budget allocation and team mobilization.

Cross-Functional Alignment Rate: Percentage of major strategy decisions where marketing, sales, product, and finance use the same data instead of conflicting reports. Insight-driven organizations achieve 85% or higher alignment on forward-looking decisions. Measure by auditing strategy meetings for shared or siloed data, and track decisions that need realignment due to conflicting assumptions.

Predictive Accuracy: Measure how closely forecasted outcomes match actual performance for strategic bets made 60-90 days ahead. Unlike short-term campaign predictions, this metric tracks leaders' ability to anticipate market reactions to competitors' launches, positioning changes, and other market developments. Track accuracy each quarter. Analyze where results differ to improve scenario models.

Opportunity Capture Rate: Percentage of validated market opportunities acted upon within optimal timing windows before competitive response diminishes advantage. Calculate by documenting identified opportunities, measuring time-to-action, and tracking whether late entry reduced potential returns. Leaders who consistently achieve capture rates of 70% or higher typically outperform their peers by 15-25% in market share growth.

These metrics drive accountability for learning, not just quarterly results. Leadership teams use them to strengthen competitive advantage with faster and more accurate market responses.

Action Steps Implemented

- **Created an "Early Signals" automated alert system** with specific triggers: 15%+ week-over-week change in consumer sentiment, 25%+ shift in partner sales velocity by category, competitor activity detected through social monitoring, and regulatory changes affecting product categories. Alerts are routed to the CEO, CMO, and Head of Product, with a 24-hour response requirement for high-priority signals.

- **Established a 14-day decision window protocol** for high-confidence opportunities, defined as signals meeting three criteria: cross-channel validation (minimum two data sources confirming the trend), demographic alignment with core customer segments, and projected ROI exceeding 15% within 90 days. Medium-confidence signals are subject to 30-day evaluation periods with designated decision checkpoints.

- **Instituted monthly "Future Readiness" leadership reviews** alongside standard financial reporting, focusing on forward-looking metrics rather than lagging performance indicators. Sessions include competitive intelligence updates, analysis of emerging consumer behavior, and discussions on resource allocation for identifying nascent opportunities.

- **Partnered with three key distributors** to receive weekly sell-through data and monthly consumer feedback summaries, reducing signal detection lag from quarterly reports to real-time insights. Established formal data-sharing agreements with standardized reporting formats and mutual early-warning protocols for market shifts.

- **Scheduled semi-annual board workshops** dedicated to reviewing insight-driven strategic pivots, documenting lessons learned from both successful accelerations and strategic redirections. These sessions enhance institutional knowledge of signal interpretation accuracy and inform future decision-making frameworks.

- **Implemented cross-functional boardroom protocols** activated when opportunity capture windows fall below 21 days, enabling rapid resource reallocation and decision authority delegation to front-line teams without standard approval hierarchies.

Cross-Functional Implementation: Strategic Intelligence Accountability Matrix

Predictive leadership requires coordination across every function that detects, interprets, and acts on market signals. Unlike traditional leadership approaches that rely on departmental reports, insight-driven leadership embeds signal detection throughout the organization while maintaining central interpretation and decision authority. As a result, when emerging opportunities require rapid response, executive teams must coordinate cross-functional intelligence without sacrificing decision speed or strategic focus. To address this need, the following accountability matrix ensures that strategic intelligence becomes an organizational capability rather than an individual intuition.

- **Executive Leadership:** Owns signal interpretation, scenario modeling, and strategic decision authority. Maintains cross-functional alignment and resource allocation protocols for time-sensitive opportunities.

- **Data Analytics:** Provides predictive modeling, signal detection algorithms, and competitive intelligence systems. Creates executive dashboards and maintains early warning systems across all business functions.

- **Strategy Operations:** Manages decision workflow acceleration, cross-functional communication protocols, and strategic initiative tracking. Ensures rapid organizational mobilization when opportunities require fast action.

- **Business Intelligence:** Monitors competitive responses, market validation, and strategic outcome measurement. Provides post-decision analysis and strategic learning integration.

- **Department Liaisons:** Provide functional expertise, implementation feasibility assessment, and resource availability updates. Ensure strategic decisions can be executed effectively within existing operational constraints.

Executive Reflection – Horizon Scanning:

Consider this: If a competitor were to launch into your category within the next 90 days, what signals would indicate their entry, and would your leadership team respond swiftly enough to meet them in the market?

Chapter 16 Checklist: Future-Ready Leadership

- Monitor leading indicators in real-time, rather than relying solely on trailing sales reports.

- Align cross-functional strategy meetings around shared insight dashboards.

- Set decision windows to act on emerging opportunities.

- Track and improve predictive accuracy rates on a quarterly basis.

- Formalize a partner data-sharing program to expand early signal coverage.

- Establish automated alert triggers for market shifts exceeding predetermined thresholds.
- Define high-confidence signal criteria and corresponding decision protocols.
- Create fast-track budget allocation processes for time-sensitive opportunities.
- Document failed predictions to improve scenario modeling accuracy.
- Schedule regular "horizon scanning" sessions to identify weak but emerging market patterns.

By implementing these frameworks, Northbridge transformed from a reactive organization that responded to market changes into a predictive one that shapes them. The key benefits have been greater agility, proactive decision-making, and the ability to anticipate market needs. But sustainable insight-driven leadership requires more than individual success stories; it demands organizational maturity that scales beyond any single leader or department.

The challenge every executive faces: How do you assess your organization's ability to capture, interpret, and act on first-party insights for faster market response, sharper forecasts, and reduced risk? How do you pinpoint gaps that slow decision cycles and hinder market predictions? Most importantly, how do you craft a roadmap that turns data goals into measurable competitive advantages?

Without clear benchmarks for data maturity, ambitious visions stall in execution, resulting in slower reactions, missed opportunities, and wasted resources. Teams struggle with conflicting priorities, departments operate from different data sources, and leadership lacks visibility into which capabilities need development first.

The next chapter introduces the First-Party Data Maturity Model—a systematic framework that turns abstract "data-driven" goals into concrete, measurable progress across every department. With this model, organizations can achieve faster decision-making, improved customer targeting, and ongoing innovation by diagnosing their current state, prioritizing capability investments, and tracking progress toward insight-driven market leadership with the same rigor they apply to financial performance.

Because the companies that master this progression don't just implement better analytics, they unlock durable strategic advantages, such as higher win rates, stronger customer loyalty, and market agility, building sustainable competitive moats that compound with every market cycle. Take deliberate steps now to advance along this progression and secure these compounding advantages for your organization.

Chapter Seventeen

The First Party Data Maturity Model:

Transforming Data Collection Into Strategic Assets

Organizational data maturity transforms from ad hoc information gathering into systematic competitive intelligence through structured capability development, measurable progression frameworks, and disciplined advancement protocols. This chapter presents the First-Party Data Maturity Model, which transforms abstract "data-driven" aspirations into concrete advancement roadmaps through systematic assessment and targeted capability investments.

We'll examine five maturity stages that convert data collection into strategic intelligence, introduce assessment frameworks that identify specific advancement barriers, and provide progression protocols that ensure consistent maturity velocity—the speed at which organizational capabilities advance through measurable competency levels. You'll learn to establish measurement systems that track capability gaps and readiness for advancement across all business functions.

By the end of this chapter, you'll have operational blueprints for assessing current data maturity, advancement frameworks that maintain realistic timelines while scaling organizational capabilities, and measurement systems that prove maturity ROI while accelerating competitive positioning and market intelligence capabilities.

Northbridge: From Data Collection to Market Intelligence

Wednesday morning workshop. The boardroom lights dim as a five-step graphic glows on the wall. From left to right, the bars rise in both height and depth of color, representing the stages of awareness: ***Blind, Reactive, Aware, Strategic,*** *and* ***Intelligent.*** Evelyn faces the team.

"This is the First-Party Data Maturity Model," she says, pointing to the first step. "Every organization is somewhere on this staircase, from running blind on third-party scraps to fueling growth with real-time, predictive insights. Our job is to know exactly where we stand today, and what it will take to move to the next step."

Northbridge's climb up the model followed a clear arc: moving from **Reactive** to **Aware** by centralizing data created a single source of truth; advancing to **Strategic** made first-party data integral to marketing, sales and product decisions, which reduced customer acquisition cost by 18%, raised campaign conversions by 25% and improved cross-sell success by 35%; reaching **Intelligent** added real-time decisioning and predictive insight, delivering 23% faster time-to-market, 31% higher customer lifetime value, and 40% more accurate demand forecasts. Taken together, advancing three full stages over several years produced 15-20% revenue growth above historical performance. This is not only an operational shift; but also a competitive necessity.

The room is quiet. Each stage on the chart tells a story:

- **Blind** – No structured data collection, just hunches and disconnected reports.

- **Reactive** – First-party data is gathered, but trapped in silos.

- **Aware** – Data is connected across departments, finally providing a shared view.

- **Strategic** – Insights shape marketing, sales, and product priorities.

- **Intelligent** – Real-time data drives every decision, with predictive models steering growth.

Northbridge has made big gains in the past year, but Evelyn knows that without a clear path, progress will stall. This is where the maturity model comes in. It isn't just a chart; it's the roadmap for how every team will climb from where they are now to where they need to be. *Figure 3*

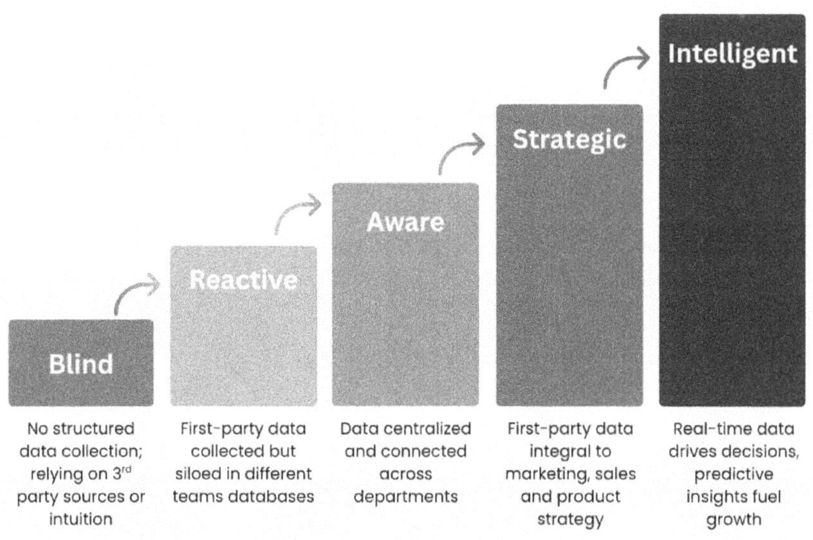
The First-Party Data Maturity Model
Blind
Reactive
Aware
Strategic
Intelligent
No structured data collection; relying on 3rd party sources or intuition
First-party data collected but siloed in different teams databases
Data centralized and connected across departments
First-party data integral to marketing, sales and product strategy
Real-time data drives decisions, predictive insights fuel growth

Figure 3: First-Party Data Maturity Model

The Five Stages: A shared language for progress:

Organizations rarely transition from data darkness to real-time intelligence overnight. Progress comes in deliberate steps. The table on the next page outlines the five steps of Northbridge's model, complete with hallmark traits, common risks, and a current snapshot of the company's standing today.

Why this matters:

Without a shared definition of progress, each department risks using its own standards of maturity, which can lead to inconsistent results. The model establishes a single scoreboard, ensuring that everyone measures success in the same way.

Name	Hallmark Traits	Primary Risk	Northbridge Example
Blind	Decisions driven by intuition, siloed reports, and minimal capture	Wasted spend	Northbridge pre-project: sampling events with no first-party validation
Reactive	Basic capture, siloed systems, lagging reports	Slow course correction	Early QR pilots without unified IDs
Aware	Unified profiles, descriptive dashboards, and some cohort insight	Insight latency	Current state: weekly Flywheel cadence, Holdout tests
Strategic	Predictive models, cross-functional KPIs, proactive actions	Complexity management	Goal in 6-12 months
Intelligent	Real-time automation, AI optimization, continuous experiments	Over-reliance on automation	Long-term vision

Five Stages

Implementation Timeline Reality Check

Maturity advancement cannot be rushed. Organizations typically require 6-9 months to progress one full stage, with the *Aware* to *Strategic* transition often taking 12-18 months due to the complexity of predictive modeling and challenges with cross-functional alignment. Organizations should budget 3-4 months for infrastructure, 2-3 months for process development, and 2-3 months for validation and optimization at each stage. Attempting to skip stages or compress timelines beyond organizational capacity results in failed implementations, team frustration, and leadership skepticism about data initiatives.

The progression barriers are predictable. Moving from *Blind* to *Reactive* requires foundational infrastructure investment, data capture systems, consent management, and basic reporting capabilities. *Reactive* to *Aware* demands the more complex work of identity unification and cross-departmental data sharing protocols. Strategic advancement requires sophisticated analytics capabilities and effective organizational change management as teams transition from intuition-based to model-driven decision-making.

Most failures happen when organizations underestimate the required governance and training at each stage. Technical infrastructure accounts for only 40% of the effort; most advancement requires redesigning processes, developing skills, and adapting the culture to support insight-driven operations.

Failure Examples: When Advancement Goes Wrong

The Predictive Model Trap: Northbridge's first attempt at Strategic advancement in 2023 failed spectacularly. Eager to implement churn prediction models, the marketing team built algorithms using Reactive-level data quality, incomplete customer profiles, inconsistent tracking, and siloed departmental databases. The resulting models generated a 73% false positive rate, triggering retention campaigns for customers who were already planning to increase their spending. Leadership confidence in data-driven decisions eroded over a six-month period.

The Governance Shortcut: A competitor attempted to accelerate from Aware to Intelligent within six months by implementing real-time personalization without Strategic-level governance frameworks. A data breach affecting 40,000 customer records during the pilot phase resulted in $2.3M in regulatory fines and necessitated a complete system rollback. Their maturity progression was set back 18 months while rebuilding compliance infrastructure.

The Silo Skip: Another organization tried to bypass cross-functional alignment by having marketing achieve Strategic capabilities independently. When predictive models recommended campaign strategies that conflicted with sales priorities and product roadmaps, execution became fragmented. Campaign performance declined 15% due to mixed messaging and resource conflicts, forcing a return to coordinated Aware-level operations.

These failures reinforce a fundamental principle: sustainable maturity advancement requires disciplined progression through each stage's foundational requirements before attempting the next level's capabilities.

Enhanced Self-Assessment Scoring Framework

Organizations should tailor their scoring criteria to their specific industry context and strategic priorities. Use this framework to develop stage-appropriate benchmarks:

Scoring Scale Definition:

- **Level 1:** Foundational gaps exist that prevent basic functionality
- **Level 2:** Basic capability present but inconsistently applied
- **Level 3:** Systematic processes established with documented procedures
- **Level 4:** Advanced capabilities with predictive elements and automation
- **Level 5:** Industry-leading performance with continuous optimization

Customization Guidelines by Dimension:

Capture Scoring Framework

What You're Measuring: The breadth and quality of first-party data collection across customer touchpoints. This evaluates both the percentage of interactions that capture consented data and the sophistication of collection methods.

- **Level 1:** Less than 30% customer touchpoints capture consented data
- **Level 2:** 30-60% touchpoint coverage with manual data entry
- **Level 3:** 60-85% coverage with standardized capture protocols
- **Level 4:** 85-95% coverage with real-time validation and enrichment

- **Level 5:** 95%+ coverage with predictive data quality monitoring

Analysis Scoring Framework:

What You're Measuring: The sophistication of insights generated from collected data. This assesses whether your organization can transform raw data into actionable intelligence that drives business decisions.

- **Level 1:** Reporting limited to basic demographic summaries
- **Level 2:** Descriptive analytics with trend identification capability
- **Level 3:** Cohort analysis and customer journey mapping
- **Level 4:** Predictive modeling with 70%+ accuracy rates
- **Level 5:** Real-time machine learning optimization with continuous model improvement

Activation Scoring Framework:

What You're Measuring: The speed and sophistication of acting on insights. This evaluates how quickly your organization can translate analysis into customer-facing actions and campaign optimizations.

- **Level 1:** Manual campaign execution with weekly decision cycles
- **Level 2:** Scheduled automation with pre-defined audience segments
- **Level 3:** Dynamic segmentation with same-day campaign adjustments
- **Level 4:** Real-time personalization with behavioral triggers
- **Level 5:** Predictive activation with micro-moment optimization

Governance Scoring Framework:

What You're Measuring: The robustness of privacy, compliance, and data stewardship practices. This assesses whether your data operations meet regulatory requirements and maintain customer trust at scale.

- **Level 1**: Basic privacy policy without enforcement mechanisms
- **Level 2:** Documented data retention with manual compliance checking
- **Level 3:** Automated consent management with audit trails
- **Level 4:** Continuous compliance monitoring with breach detection
- **Level 5:** Predictive governance with proactive risk mitigation

Velocity Scoring Framework:

What You're Measuring: The speed at which insights translate into strategic and tactical decisions. This evaluates organizational learning cycles and the time from signal detection to action implementation.

- **Level 1:** Insights take 30+ days to influence strategic decisions
- **Level 2:** 15-30 day insight-to-action cycles with quarterly reviews
- **Level 3:** Weekly insight integration with monthly strategy adjustments
- **Level 4:** Daily insight updates with real-time tactical optimization
- **Level 5:** Continuous learning loops with automated strategy refinement

The Northbridge results are displayed in a heatmap on the executive dashboard, a visual truth serum that reveals where strengths and gaps truly lie.

Department	Capture	Analysis	Activation	Governance	Velocity
Marketing	4	4	4	3	3
Sales	3	3	3	3	2
Product	3	3	2	3	2
Finance	2	2	2	4	2

Northbridge Self Assessment

This quarter's read:

- Marketing leads the pack but needs better governance discipline.
- Sales show strong activation but slow learning velocity.
- Finance captures selectively but needs deeper analysis capabilities.

The assessment revealed critical insights: Marketing's governance gap (Level 3) was blocking their progression to the Strategic stage, despite strong capture and analysis capabilities (Level 4). Sales velocity scores (Level 2) indicated that they had strong activation but couldn't learn quickly enough from outcomes. Most surprisingly, Finance's governance leadership (Level 4) positioned them to enable organization-wide advancement.

The 18-month roadmap prioritized governance infrastructure first. Marketing invested $45K in consent management automation, achieving Level 4 governance within 90 days and unlocking Strategic-stage predictive modeling. Sales focused on velocity improvement through automated learning loops, reducing insight-to-action cycles from 15 days to 3 days within one quarter.

Climbing the Steps

1. **From Blind to Reactive** – Introduce foundational capture: QR codes, links, receipt validation, consent ledger.

2. **Reactive to Aware** – Unify IDs, build descriptive dashboards, establish the Flywheel cadence.

3. **Aware to Strategic** – Launch predictive models, link insights directly to budgeting and product roadmaps.

4. **Strategic to Intelligent** – Automate decision loops, embed AI optimization, maintain human oversight.

Northbridge Roadmap Milestones

- **Q1: Foundation Consolidation** - Achieve 90%+ unified ID match rate across marketing and sales touchpoints, completing the transition to the Aware stage for core customer-facing departments. **Risk Factor:** The complexity of legacy CRM integration may extend unified ID deployment into early Q2. **Mitigation:** Parallel data warehouse upgrade approved to accelerate matching algorithms.

- **Q2: Predictive Model Deployment** - Launch churn propensity and expansion likelihood models with accuracy rates exceeding 70%, advancing marketing and sales toward the Strategic stage. **Dependencies:** Q1 data quality targets must be met before model training begins. **Success Metrics:** 15% improvement in retention campaign effectiveness, 25% increase in expansion opportunity identification.

- **Q3: Real-Time Activation Engine** - Implement automated offer optimization for the top 20% customer cohorts, reducing response latency from hours to minutes. **Investment Required:** $180K in marketing automation platform upgrades and API integrations. **Expected ROI:** 12% lift in campaign conversion rates within 90 days.

- **Q4: Intelligent Marketing Achievement** - Reach the Intelligent stage in marketing activation while establishing Strategic-level capabilities company-wide. **Validation Criteria:** Sub-5-minute trigger response times, predictive model precision maintaining >75% accuracy, and cross-functional KPI alignment achieving 85%+ shared data source utilization.

Governance that Scales with Capability

Northbridge advances data governance in lockstep with technical capability. A cross-functional Data Steward Council reviews monthly service-level objectives (SLOs) to keep risk low and trust high:

- **Permission state coverage:** ≥99% of active profiles have a documented **lawful basis and channel-level preferences** (e.g., email opt-out honored, SMS consent captured), with auditable provenance. *Note:* lawful bases and consent requirements vary by channel and jurisdiction; SMS generally requires prior express consent, while email marketing in the U.S. is governed by CAN-SPAM's opt-out regime. [1]

- **Critical-field quality:** ≥95–98% completeness & validity on priority entities (customers, products, stores), measured against recognized data-quality dimensions (e.g., ISO/IEC 25012 data quality model).

- **Incident response: Internal** breach-escalation SLA <12 hours; **external** notifications follow applicable law.

- **Regulatory alignment:** Quarterly internal audits with remediation of any findings; aim for zero critical open items each quarter.

---

1. NIST. National Institute of Standards and Technology. US Department of Commerce. Artificial Intelligence Risk Management Framework (AI RMF 1.0)

Maturity Gates (Northbridge Model)

Teams don't progress without meeting governance thresholds for their current stage. We use a pragmatic, adapted maturity model (informed by DAMA/Gartner) and make the requirements explicit:

- **Reactive** → basic lawful-basis capture, retention policy, data catalog entries for priority tables.
- **Aware** → cross-department data-sharing agreements, DPIAs/PIAs for new uses.
- **Strategic** → model-risk governance: **bias monitoring**, explainability, decision logs.
- **Intelligent** → near-real-time policy enforcement, automated consent propagation, continuous controls testing.

Budgeting for Governance

There is **no universal benchmark** for governance as a % of data platform spend. As planning guidance, we recommend reserving ~10–20% (rising with the scope of risk and automation) for stewardship, controls, monitoring, audits, and reporting, and revisiting this allocation annually. The key is not the exact % but ensuring controls, auditability, and model governance keep pace with capability.

Key Metrics for Progression, Gatekeepers to the next step:

Progression in the model isn't subjective. The scoreboard on the next page lists the quantitative thresholds that Northbridge must meet before completing a stage: unified match rate for *Aware*, model precision for *Strategic*, trigger latency for *Intelligent*, and learning velocity. Use these targets to decide where to invest resources.

Metric	Stage Trigger	Target	Why This Matters	Measurement Method
Unified contact match rate	Aware threshold	90%+ deduplicated	Enables accurate customer journey tracking and prevents campaign overlap	Monthly identity resolution audit across all customer databases
Predictive model precision	Strategic threshold	≥ 0.70 AUC for churn model	Ensures model-driven decisions outperform intuition-based approaches	Quarterly model performance validation using holdout test sets
Real-time trigger latency	Intelligent threshold	< 5 minutes from event to action	Captures micro-moments when customer intent is highest	Automated monitoring of event processing and campaign activation pipelines
Learning velocity improvement	Cross-stage	Reduce by 20% each quarter	Accelerates organizational adaptation to market changes	Time-to-insight tracking from signal detection to strategic decision implementation
Cross-functional data alignment	All stages	85%+ shared source utilization	Prevents conflicting strategies and resource waste	Monthly audit of departmental reporting sources and KPI calculation methods

Key Metrics For Progression

Sidebar – Real-world example

Today, Promolytics tracks maturity progress by capturing the percentage of campaigns that move from raw data to activated insight. At Northbridge, early tracking revealed that while capture and analysis were strong, activation speed lagged. This visibility let the team target specific process bottlenecks instead of guessing where improvements were needed.

Future AI and ML features could automatically flag when a department's maturity score risks slipping, prompting corrective action before performance dips.

Action Steps Implemented

- Added the maturity model heatmap to the executive dashboard.
- Linked capital budget approvals to maturity targets (e.g., data warehouse upgrade unlocks *Strategic* stage).
- Established quarterly data review workshops.
- Created an incentive program: next-stage achievers earn discretionary innovation budgets.
- Established department-specific advancement timelines with milestone dependencies.
- Created maturity gate criteria that prevent premature stage progression.
- Implemented monthly cross-functional maturity review sessions.
- Developed governance investment budgets tied to maturity advancement goals.

Cross-Functional Implementation: Maturity Advancement Accountability Matrix

Data maturity advancement requires coordination across every function that generates, analyzes, or acts on customer intelligence. Unlike technology-focused data initiatives, systematic maturity progression embeds capability development throughout the organization while maintaining consistent advancement standards. When maturity gaps threaten competitive positioning, cross-functional teams must coordinate assessment, development, and validation activities. The following accountability matrix ensures that maturity advancement becomes organizational capability rather than departmental initiative.

- **Data Governance Council:** Owns maturity assessment standards, advancement gate criteria, and cross-functional alignment protocols. Maintains progression oversight and resource allocation decisions.

- **Department Maturity Leads:** Provide functional assessment, capability development planning, and advancement validation. Ensure departmental readiness for next-stage requirements and integration protocols.

- **Technology Integration:** Manages infrastructure requirements, system capabilities, and technical advancements to enable effective use. Provides platform readiness assessment and integration support.

- **Change Management:** Coordinates training programs, process adaptation, and cultural alignment initiatives. Ensures organizational readiness for capability advancement and sustainable adoption.

- **Performance Analytics:** Monitors advancement metrics, tracks capability development, and validates progression achievements. Provides objective maturity measurement and verification of advancement.

Executive Action – Maturity Assessment:

Based on your department heatmap scores, which capability gap represents the greatest bottleneck to advancing one full maturity stage within the next 12 months? Consider: Will improving data quality unlock better analysis, or does governance lag pose compliance risks that could force regression? Does your team have the organizational change management capacity to implement predictive models, or should you focus on perfecting current-stage capabilities before advancement?

Follow-up Question: What's the minimum investment threshold (time, budget, and personnel) required to achieve measurable progress on your identified bottleneck within 90 days, and do you have executive commitment to sustain that investment through inevitable implementation challenges?

Chapter 17 Checklist: Advancing on the Maturity Model

- Complete a comprehensive department self-assessment using the five-dimensional scoring framework.
- Identify the lowest-scoring capability dimension and define a 90-day improvement plan with specific milestones.
- Establish realistic advancement timelines based on organizational capacity and resource constraints.
- Add maturity stage progression KPIs to each department's quarterly OKRs and performance reviews.
- Schedule monthly Data Stewardship Council reviews tied directly to stage advancement criteria.
- Document governance requirements for current and target maturity stages.
- Reserve innovation pilot budgets unlocked by successful stage advancement achievements.
- Create cross-functional alignment protocols, ensuring shared data sources across departments.
- Implement automated governance monitoring appropriate for the current maturity stage.
- Establish failure learning processes to capture insights from unsuccessful

advancement attempts.

By the end of the workshop, every Northbridge department had a clear understanding of its exact position on the maturity model, as well as the specific capabilities holding it back from advancing. The heatmap was used not only as a scorecard but also as a strategic prioritization tool. Key takeaways included the realization that governance improvements, more than increased analytics investments, would drive faster progress.

Evelyn closed the session with a critical reminder: "Maturity models only create value when they drive action. Progress doesn't happen because we measure ourselves—it happens because we systematically address the gaps those measurements reveal."

The path forward required two foundational commitments: First, accepting that sustainable advancement takes time and cannot be rushed without risking costly setbacks. Second, recognizing that each stage builds essential capabilities for the next, attempting shortcuts inevitably forces organizations backward when foundational gaps surface under advanced use cases.

The ultimate goal isn't just reaching Intelligent status; it's building organizational capabilities that consistently enhance competitive advantage every quarter. The key takeaway: companies that excel at this disciplined progression don't just improve analytics—they create institutional learning systems that adapt more quickly than competitors and proactively anticipate market shifts, while others are still analyzing past results.

The next challenge transforms this maturity roadmap from a planning document into an operational reality. Success requires more than frameworks and metrics; it demands systematic change management that aligns teams, processes, and incentives around insight-driven growth. The following chapter addresses the leadership and cultural shifts necessary to sustain maturity advancement and capture the full competitive potential of first-party intelligence.

Chapter Eighteen

The Foundation Framework:

When Infrastructure Development Prepares Organizations for Intelligence

Data foundation building transforms scattered information collection into a systematic organizational capability through the development of structured infrastructure, the implementation of coordinated processes, and the measurement of capability assessment. This chapter provides the first-quarter blueprint that establishes the foundational systems and governance frameworks necessary to advance through higher maturity stages.

We'll examine foundational infrastructure requirements that enable future predictive capabilities, introduce assessment protocols that identify current capability gaps, and provide implementation frameworks that ensure realistic foundation velocity—the speed at which basic data systems can be established without overwhelming organizational capacity. You'll learn to establish baseline measurement systems and governance protocols that prepare your organization for the systematic advancement of maturity outlined in previous chapters.

By the end of this chapter, you'll have operational blueprints for establishing data foundations, realistic implementation timelines based on organizational resources and constraints, and measurement systems that validate infrastructure readiness for advancing toward higher maturity stages. This foundation work typically requires 6-18 months before organizations can effectively pursue the complete strategic and intelligent capabilities demonstrated throughout this book, though timelines may be shorter for companies implementing selective components rather than the full transformation.

Northbridge: Foundation First

Thursday morning, boardroom. The maturity assessment heatmap from yesterday's workshop dominates the main screen, revealing Northbridge's current capabilities and the requirements for advancement. Evelyn stands beside a quarterly planning board, marker in hand.

"We're not launching campaigns today," she says, addressing the leadership team. "We're building the infrastructure that will power campaigns for the next two years. The assessment showed us where we are—now we design the foundation that gets us where we need to be."

She points to the heatmap, which shows marketing at Level 3-4 capabilities but governance gaps, sales with strong activation but weak velocity, and finance with selective capture but analysis limitations. "Foundation first means we fix the infrastructure before we build advanced capabilities on unstable ground."

The room understands: this isn't about quick wins or immediate campaign optimization. This is about systematic capability building that enables the strategic and intelligent maturity stages the organization needs to reach.

"The next ninety days aren't about transformation," Evelyn continues. "They're about establishing the systems, processes, and governance that make transformation possible. Strategy is only as strong as its foundation."

Foundation Readiness Assessment

Before beginning foundation building, organizations must honestly assess their starting capabilities and resource constraints:

Infrastructure Audit:

- Current data capture coverage across customer touchpoints

- Existing consent management and privacy compliance status
- Identity resolution capabilities and data quality standards
- Analytics infrastructure and reporting systems

Organizational Readiness:

- Cross-functional leadership commitment and availability
- Technical resource allocation and skill assessment
- Change management capacity and training requirements
- Budget allocation for foundational improvements

Timeline Reality Check:

- Foundation building requires sustained effort over quarters, not weeks
- Organizations typically need 6-12 months for basic infrastructure
- Advanced capabilities require an additional 6-12 months after foundation completion
- Resource constraints may extend timelines, but should not compromise quality

The Foundation Building Framework

Days 1-30: Assessment and Planning

- Complete a comprehensive maturity assessment using the Chapter 16 framework
- Establish a data governance council with clear roles and meeting cadence
- Audit all customer touchpoints for data capture opportunities
- Document current consent management and privacy compliance gaps
- Define unified customer identity requirements and technical specifications
- Create a realistic timeline based on organizational capacity and resources

Days 31-60: Infrastructure Development

- Implement basic consent management and preference center functionality
- Begin unified ID resolution pilot across primary marketing touchpoints
- Establish data quality monitoring and improvement protocols
- Launch foundational analytics dashboards for key performance metrics
- Train teams on new processes and establish governance protocols

- Document lessons learned and refine the implementation approach

Days 61-90: Validation and Optimization

- Deploy a pilot precision marketing campaign using new infrastructure
- Measure baseline improvements in data quality and campaign performance
- Validate governance protocols and privacy compliance procedures
- Conduct a cross-functional review of foundation building progress
- Plan next quarter's advancement toward higher maturity capabilities
- Document the foundation-building playbook for other organizations or departments

Cross-Functional Implementation: Foundation Building Accountability Matrix

Foundation building requires coordination across every function while maintaining realistic expectations about capability development timelines. Unlike advanced maturity stages that require sophisticated predictive capabilities, foundation building focuses on establishing the infrastructure and processes that enable systematic advancement. The following accountability matrix ensures foundation building becomes a coordinated organizational effort rather than a departmental initiative.

- **Data Governance Council:** Owns overall foundation strategy, cross-functional coordination, and advancement gate criteria. Maintains realistic timeline expectations and resource allocation decisions.

- **Marketing Operations:** Implements consent management systems, establishes basic audience segmentation, and creates baseline measurement protocols. Focuses on data quality improvement rather than advanced personalization.

- **Data Engineering:** Develops unified ID resolution capabilities, implements data capture improvements, and establishes analytics infrastructure. Prepares technical foundation for future predictive capabilities.

- **Analytics:** Conducts capability assessments, establishes baseline performance metrics, and develops a roadmap for advanced analytics. Focuses on descriptive analytics excellence before predictive modeling.

- **Compliance/Legal:** Ensures privacy compliance, develops governance frameworks, and establishes audit protocols. Creates a scalable compliance infrastructure to support future capability advancements.

Foundation Metrics That Matter

Track foundation building progress with metrics that indicate infrastructure readiness rather than campaign performance:

- **Data Quality Score:** Percentage of customer records with complete, accurate core attributes

- **Consent Coverage Rate:** Percentage of customer touchpoints with proper consent collection

- **Identity Resolution Rate:** Percentage of customer interactions successfully linked to unified profiles

- **Process Compliance Rate:** Percentage of data handling activities following established governance protocols

- **Cross-Functional Alignment Score:** Percentage of major decisions using shared data sources rather than conflicting reports
- **Foundation Readiness Index:** Composite score indicating organizational preparedness for the next maturity stage advancement

Realistic Timeline Expectations

Quarter 1 (Days 1-90): Foundation Building

- Establish basic infrastructure, governance, and measurement systems.

Quarters 2-3 (Days 91-270): Foundation Optimization

- Refine processes, improve data quality, and prepare for advanced capabilities.

Quarters 4-6 (Days 271-540): Strategic Advancement

- Begin implementing predictive capabilities and integrating strategic decisions.

Beyond 18 Months: Intelligent Capabilities

- Deploy real-time optimization and fully automated decision systems.

Organizations attempting to compress these timelines risk the failures documented in Chapter 17, including poor data quality leading to inaccurate models, governance shortcuts that create compliance vulnerabilities, and cross-functional misalignment that causes execution fragmentation.

Sidebar – Real-world example

Promolytics gives organizations a paced, auditable way to build their data foundation before pursuing advanced capabilities. Out of the box it unifies data capture across QR flows, surveys, and coupons or receipt uploads, then ties each event to a consented profile with clear source attribution.

Single source shared data with a simple event dictionary standardizes what is collected and how it is labeled, which reduces integration friction between marketing, sales, product development, analytics, and IT.

In practice this moves a company from **Reactive** too **Aware** by centralizing events and identities, then toward **Strategic** by making shared definitions and readiness metrics visible to all leaders.

When the foundation holds steady, Promolytics provides a clear runway to layer predictive scoring and real-time activation with lower risk.

Executive Reflection – Execution Discipline:

To begin, determine which milestone in the next thirty days lacks a clear owner or metric. Next, identify the immediate adjustments required to secure delivery.

Chapter 18 Checklist: Building Your Data Infrastructure

- Publish the 30-60-90 roadmap to all department heads. Ensure each head reviews and acknowledges their section. Confirm that department heads share the roadmap with their teams within 48 hours.

- Assign owners and due dates to every weekly milestone. Immediately confirm assignments with each owner and set calendar reminders for all deadlines to ensure timely completion.

- Activate dashboards for match rate, churn, and lift tracking. Check access for all key stakeholders and schedule a dashboard walk-through within the first week.

- Schedule executive stand-ups and Flywheel Reviews through Day 90. Send calendar invites to all participants now and confirm attendance within 24 hours.

- Allocate contingency resources for identified risks. Notify key leads of available backup resources and confirm understanding of escalation protocols.

The wall calendar is now filled with color-coded milestones, each with an owner, a deadline, and a metric. Still, Evelyn knows that execution doesn't live in a Gantt chart — it lives in the habits, meetings, and daily decisions of each team. The ninety-day plan may set the pace, yet without clear playbooks for every department, momentum will fade and silos will re-emerge. This shift from the plan to real-world execution raises a new challenge: sustaining momentum across teams.

This is why the next step is to break the strategy down into department-specific operating guides, ensuring that marketing, sales, product, operations, and governance all move in lockstep — not just toward their own goals, but toward the shared maturity model Northbridge is climbing.

Chapter Nineteen

Department Playbooks:

Operational Guidelines for Cross-Functional Execution

Department execution transforms from ad hoc project management into systematic operational excellence through coordinated playbooks, unified metrics, and disciplined accountability frameworks. This chapter outlines the departmental implementation system, which transforms strategic insights into repeatable daily actions through structured ownership protocols and cross-functional alignment mechanisms.

We'll examine five departmental playbooks that convert the INSIGHT framework into specific weekly actions, introduce accountability systems that ensure consistent execution velocity, and provide escalation protocols that maintain operational momentum when bottlenecks emerge. You'll learn to establish operational rhythms that track departmental performance while maintaining alignment with organizational maturity advancement goals.

By the end of this chapter, you'll have operational playbooks for implementing systematic departmental execution, accountability frameworks that maintain coordination while scaling individual department capabilities, and measurement systems that prove operational ROI while accelerating cross-functional collaboration and achieving strategic objectives.

Northbridge: From Strategy to Systems

Friday afternoon. With the ninety-day plan underway, Evelyn shifts the focus from *what* to do to *how* to do it consistently. She meets each department head to turn strategic goals into routine, repeatable actions.

"Momentum matters more than speed," she reminds them. "A strong quarter isn't built on heroic sprints, it's built on disciplined actions repeated week after week, with the right metrics to keep us honest."

The result: **department playbooks** mapped to the INSIGHT framework. These aren't one-off project plans. They are systems helping marketing, sales, product, operations, and governance teams use data to drive revenue—without confusing data capture with value.

Each playbook is designed to:

- Establish **weekly routines** and dashboards to ensure nothing is missed.
- Outline the **Top Five Actions** for each INSIGHT stage, ensuring every handoff is clear.
- Anchor decisions to **Primary KPIs** so success is measured, not assumed.
- Define **Escalation Rules** so bottlenecks and risks are resolved fast.
- Set **Quarterly Improvement Targets** to push the organization further up the Maturity Model.

These playbooks give every department a **clear path and keep** everyone connected. Gaps become visible, dependencies clear, and the organization moves forward in sync.

How to read the following playbooks:

Each department follows a clear pattern: the Core Mission, Weekly Rhythm, INSIGHT Stage Ownership, Escalation Triggers, and Success Metrics. Treat each as a "mini-SLA." If any action, or owner is blank in your organization, immediately assign responsibility or fill the gap. Review each for missing information, and resolve these before proceeding. **Turn the page to begin.**

Marketing Intelligence Playbook

Core Mission: Convert customer signals into targeted campaigns that drive incremental lift while building the Consumer Insight Collection Strategy, which fosters customer relationships and trust through every interaction, captures consent and verified purchase behavior, and records stated preferences across all touchpoints.

Weekly Rhythm:

- **Monday:** Review engagement data and plan week's activations
- **Wednesday:** Conduct Capture Pulse review and adjust targeting
- **Friday:** Analyze campaign performance and prepare insights for sales handoff

INSIGHT Stage Ownership:

- **Identify:** Marketing Ops owns QR deployment, link tracking, and Consumer Insight Collection touchpoint expansion
- **Normalize:** Analytics owns data quality, customer journey mapping, and insight validation protocols
- **Segment:** Campaign teams' own audience definition, holdout management, and preference-based targeting
- **Interpret:** Insights Lead owns story development, pattern recognition, and consumer behavior analysis
- **Guide:** Marketing Director owns budget allocation, strategic decisions,

and insight collection prioritization

- **Harness:** CRM Admin owns automation deployment, journey management, and preference center optimization
- **Track:** Analytics owns lift measurement, fatigue monitoring, and insight collection effectiveness

Escalation Triggers:

- Engagement fatigue >15% → Pause campaigns within 4 hours
- Consent violations detected → Legal notification within 2 hours
- Lift below 5% threshold → Cross-functional review within 24 hours
- Consumer insight collection gaps >20% → Touchpoint audit within 48 hours
- Trust indicators declining → Customer experience review within 24 hours

Success Metrics:

- Incremental lift >5% per major campaign
- Engagement fatigue index <10% monthly
- Learning velocity: insight-to-action within 72 hours
- Consumer insight capture rate >75% across all customer touchpoints
- Customer trust metrics (satisfaction scores, voluntary engagement rates) are trending positive

Sales Intelligence Playbook

Core Mission: Convert marketing insights into expanded retailer partnerships, promotions, displays, and shelf placement.

Weekly Rhythm:

- **Monday:** Review account performance and prioritize high-opportunity visits
- **Wednesday:** Prepare data-backed presentations for upcoming buyer meetings
- **Friday:** Update CRM with execution feedback and competitive intelligence

INSIGHT Stage Ownership:

- **Identify:** Sales Ops owns account signal monitoring and opportunity flagging
- **Normalize:** The Data Engineering liaison owns account data quality and integration
- **Segment:** Account Managers' own territory prioritization and resource allocation
- **Interpret:** Sales Analytics owns performance story development and trend analysis
- **Guide:** The Sales Director owns strategy decisions and resource deployment

- **Harness:** Field Teams' own execution and compliance monitoring
- **Track:** Sales Analytics owns expansion metrics and pipeline velocity

Escalation Triggers:

- Account performance decline >20% → Territory manager review within 24 hours
- Execution gaps >30% → Field support deployment within 48 hours
- Competitive threat detected → Strategic response within 72 hours

Success Metrics:

- Expansion success rate >20% of Tier A accounts annually
- Average days to close reduced to <45 days
- Execution gap resolution time <72 hours

Product Intelligence Playbook

Core Mission: Transform consumer signals into validated product concepts and successful launches.

Weekly Rhythm:

- **Monday:** Review consumer feedback and unmet needs signals
- **Wednesday:** Analyze concept validation and pilot performance data
- **Friday:** Update product roadmap based on market response insights

INSIGHT Stage Ownership:

- **Identify:** Product Research owns signal detection and consumer feedback aggregation
- **Normalize:** Data Science owns concept validation and performance analysis
- **Segment:** Product Marketing owns the target audience definition and positioning
- **Interpret:** Product Strategy owns innovation prioritization and resource allocation
- **Guide:** Product Director owns roadmap decisions and launch timing
- **Harness:** Product Operations owns pilot execution and scale preparation
- **Track:** Product Analytics owns adoption metrics and success validation

Escalation Triggers:

- Pilot performance <70% of projections → Concept review within 48 hours
- Consumer sentiment shift >25% → Market research within 72 hours
- Validation failure rate >40% → Product strategy reset within 1 week

Success Metrics:

- Repeat purchase velocity <30 days for new products
- Adoption depth >40% trial-to-regular conversion within 90 days
- Concept validation accuracy >75% pilot-to-scale success rate

Operations & Field Execution Playbook

Core Mission: Ensure flawless execution that maintains brand standards and maximizes shelf velocity.

Weekly Rhythm:

Monday: Review store compliance reports and schedule corrective actions
Wednesday: Monitor inventory levels and coordinate restocking priorities
Friday: Analyze execution quality metrics and plan improvement initiatives

INSIGHT Stage Ownership:

- **Identify:** Field Teams' own compliance monitoring and gap identification
- **Normalize:** Operations Analytics owns execution data quality and reporting
- **Segment:** Regional Managers' own territory prioritization and resource allocation
- **Interpret:** Operations Intelligence owns performance pattern analysis
- **Guide:** Operations Director owns resource deployment and standard setting
- **Harness:** Field Execution owns implementation and real-time adjustments
- **Track:** Operations Analytics owns compliance metrics and remediation tracking

Escalation Triggers:

- Store compliance <85% → Field intervention within 24 hours
- Inventory stockout detected → Replenishment within 48 hours
- Execution quality decline >15% → Regional manager review immediately

Success Metrics:

- Display compliance >90% across all monitored locations
- Out-of-stock rate <5% for priority SKUs
- Remediation time <24 hours for critical execution gaps

Governance & Compliance Playbook

Core Mission: Maintain customer trust through exemplary privacy practices and regulatory compliance.

Weekly Rhythm:

- **Monday:** Review consent metrics and privacy compliance dashboards
- **Friday:** Analyze compliance trends and prepare regulatory update briefings
- **As Needed:** Audit data handling practices and update policy documentation

INSIGHT Stage Ownership:

- **Identify:** Compliance Team owns privacy signal monitoring and risk detection
- **Normalize:** Legal Operations owns data governance and policy management
- **Segment:** Privacy Analytics owns consent analysis and preference tracking
- **Interpret:** Legal Strategy owns regulatory interpretation and impact assessment
- **Guide:** Chief Legal Officer owns policy decisions and compliance strategy
- **Harness:** Compliance Operations owns enforcement and training de-

ployment

- **Track:** Compliance Analytics owns audit results and improvement tracking

Escalation Triggers:

- Consent opt-out rate >2% monthly → Campaign review within 24 hours
- DSAR response time >72 hours → Legal escalation immediately
- Policy compliance score <95% → Department audit within 24 hours

Success Metrics:

- Consent opt-in rate >85% for new touchpoints
- DSAR response time <48 hours average
- Policy compliance score >98% across all departments

Promolytics in Practice: Turning Playbooks into Measurable Wins

With Promolytics as the shared truth, each departments playbooks are connected to real-time metrics, ensuring progress remains visible and aligned.

What We Deliver Today:

Promolytics gives marketing and sales teams tools to link in-market activity to results.

- **Campaign & Experience Tracking** – Monitor consumer participation in promotions via QR code and link-based experiences.
- **Purchase Confirmation** – Validate purchases through receipt upload, linking engagement to actual sales.
- **Data Access for All Teams** – Export and share campaign performance data to keep stakeholders informed and up-to-date.
- **Pilot-Level Insights** – Identify which promotions drive the strongest engagement and redemption rates.

What's Coming Next:

Promolytics will advance beyond tracking and reporting to embed playbook execution into daily operations.

- **Cross-Department Alignment** – Marketing, sales, and product teams working from the same live performance data.
- **Revenue-Linked Accountability** – Every milestone tied directly to lift, match rate, or churn reduction.
- **Real-Time Corrections** – Instant alerts when a metric drops, with issues flagged to the right owner and team for same-day intervention.

- **AI-Guided Actions** – Recommendations for proven corrective steps based on historical turnaround success rates.

Cross-Functional Implementation: Playbook Coordination Matrix

Departmental playbooks achieve maximum effectiveness when execution is coordinated across functions rather than operating in isolation. Unlike traditional departmental planning that creates silos, systematic playbook execution embeds cross-functional dependencies and shared accountability throughout all operations. When one department's actions trigger requirements for other teams, coordination protocols ensure seamless handoffs and aligned execution across all teams. The following accountability matrix ensures playbook execution becomes an organizational capability rather than a departmental activity.

- **Marketing Operations:** Coordinates the collection of consumer insights with Sales field feedback and Product concept validation. Provides campaign performance data for Sales presentations and Product roadmap decisions.

- **Sales Intelligence:** Coordinates account insights with Marketing targeting and Operations execution monitoring. Provides retailer feedback for Product development and Marketing strategy refinement.

- **Product Development:** Coordinates consumer signals with Marketing insight collection and Sales account intelligence. Provides innovation priorities for Marketing positioning and Sales conversation enhancement.

- **Operations Execution:** Coordinates field intelligence with Sales account management and Marketing campaign optimization. Provides high-quality data for all departmental decision-making.

- **Governance Oversight**: Coordinates compliance requirements across all departments, monitoring playbook adherence and the effectiveness of escalation resolution.

Executive Reflection – Playbook Consistency:

Which department's playbook currently shows the most significant gap between weekly actions and KPI trends, and what immediate adjustment can close that gap?

Chapter 19 Checklist: Implementing Department Playbooks

- Publish playbook tables to departmental wikis and add them to onboarding materials today.

- Add a playbook adherence review to your next department meeting. Prepare to discuss ownership of gaps and agree on next steps immediately.

- Link primary KPIs to bonus structures now, working with HR or finance as needed. Make sure all criteria are clearly communicated to the team.

- Schedule a quarterly cross-playbook alignment meeting. Identify and resolve all dependency gaps in advance—set clear ownership for each action.

- Record all escalation events and resolutions for analysis to drive continuous improvement.

With every department executing from a unified playbook, Northbridge's next challenge is to ensure that each action meets the highest standards for governance, compliance, and trust.

Chapter Twenty

The Compliance Foundation:

When Trust Infrastructure Unlocks Data Strategy

Governance infrastructure transforms from reactive compliance management into proactive trust-building through systematic privacy protocols, transparent data stewardship, and measurable accountability frameworks. This chapter provides the governance foundation that transforms regulatory requirements into competitive advantages through systematic consent management and the development of customer trust.

We'll examine governance pillars that convert privacy obligations into trust-building opportunities, introduce compliance frameworks that ensure consistent regulatory adherence, and provide accountability systems that maintain governance velocity—the speed at which privacy decisions are implemented across all customer touchpoints. You'll learn to establish measurement protocols that track the quality of consent and the effectiveness of compliance across the entire data lifecycle.

By the end of this chapter, you'll have operational blueprints for implementing systematic governance processes, compliance frameworks that maintain regulatory adherence while enhancing customer trust, and measurement systems that demonstrate governance ROI while accelerating the advancement of sustainable data strategies.

Northbridge: Trust as Infrastructure

Monday afternoon. Jasmine Lee, Northbridge's legal counsel, gathers the executive team for a cross-functional checkpoint in the boardroom. On the table: a summary of new state privacy bills and a fresh industry headline about a costly data breach.

"Insight without trust is a liability," she says.

Evelyn nods. In the *Govern* stage of the INSIGHT framework, and at the "Aware" rung of the Maturity Model, governance is no longer a back-office task; it's a core operating discipline. As Northbridge's data capabilities grow, so does the responsibility to protect customer information, honor consent, and continually prove ethical data stewardship. Without this foundation, every predictive model, every sales initiative, and every marketing campaign risks collapse under the weight of regulatory failure or lost trust.

Governance Pillars for First-Party Data

Current State (Today)

- **Transparency** – Clear notice of what data is collected and the value customers receive in return.
- **Consent Integrity** – Explicit opt-in, granular preference controls, rapid opt-out enforcement.
- **Quality and Security** – Baseline data accuracy checks, encryption, and role-based access in core systems.
- **Accountability** – Named stewards for each system, documented policies for all customer-facing touchpoints.

- **Continuous Compliance** – Monitoring key state privacy regulations and updating practices annually so company policies always align with current laws.

Next State (In Progress)

- Dynamic preference management portals that update in real time.
- Automated opt-out propagation across all downstream systems within 24 hours.
- Expanded encryption to cover all in-transit datasets and vendor touchpoints.
- Continuous compliance monitoring with quarterly reviews instead of annual reviews.
- Predictive breach detection alerts are tied to anomaly patterns in access logs.

For Northbridge, achieving the "Aware" stage of governance meant implementing systematic consent management, which improved their consent opt-in rate from 67% to 89% while reducing DSAR response time from 12 days to 3 days. This governance foundation enabled the advanced marketing and sales capabilities demonstrated in previous chapters, proving that trust infrastructure unlocks rather than restricts business growth.

US Privacy Landscape: Statute Requirements Matrix

Navigating the patchwork: State privacy laws overlap but rarely align perfectly. The following matrix distills what consumers can demand, what brands must deliver, and how Northbridge already addresses each statute. As you scan, highlight any row where your organization lacks precise control; those gaps become immediate compliance priorities.

Statute	Consumer Rights	Brand Obligations	Northbridge Control
CCPA / CPRA	Access, delete, opt out of sale, correct	Consent ledger, opt-out link	Active, reviewed quarterly
Virginia CDPA	Access, delete, opt out of sale, appeal	Data protection assessment	In place, annual review
Colorado Privacy	Access, delete, opt out, data portability	Sensitive data opt-in, DPIA	Opt-in flags implemented
Utah / Connecticut	Similar core rights	Notice, security, opt out of sale	Policies aligned, audits scheduled

US Privacy Landscape

Note: Proposed federal legislation may pre-empt state laws. Monitor congressional updates and prepare impact briefs for leadership within one week of any movement.

Compliance Lifecycle at Northbridge

1. **Design** – Legal reviews all new data capture assets before launch.

2. **Implement** – Developers embed consent flags and encryption keys.

3. **Monitor** – Privacy dashboard tracks opt-out trends and DSAR turnaround.

4. **Audit** – Quarterly internal audits and annual third-party penetration tests.

5. **Respond** – Incident response playbook with 24-hour executive notification rule.

6. **Improve** – Lessons logged, policies updated, staff retrained.

Key Governance Metrics

Proof that trust is measurable: Policies sound good, but only metrics prove they're working. The scoreboard tracks five key metrics reviewed quarterly:

- **Consent Opt-In Rate:** Percentage of consumers who explicitly agree to share their data at the point of capture, measured against total interactions.

- **DSAR Turnaround Time:** Average number of days required to fulfill Data Subject Access Requests from initial receipt to completion.

- **Security Incident Resolution Time:** Average time from detection of a security incident to confirmed remediation and closure.

- **Compliance Audit Pass Rate:** Percentage of internal and external audits completed without major findings or required corrective actions.

- **Privacy Training Completion Rate:** Percentage of staff who have completed required privacy and data protection training within the designated timeframe.

If any metric drifts beyond the target, the Data Steward Council must present a corrective plan within one week.

Metric	Target	Owner
Consent opt-in rate	≥ 85%	Marketing Ops
DSAR (data subject access request) turnaround time	≤ 30 days	Privacy Team
Security incident resolution time	≤ 72 hrs.	IT Security
Compliance audit pass rate	100%	Data Steward Council
Privacy training completion rate	100%	HR & Compliance

Key Governance Metrics

Action Steps Implemented

- Rolled out yearly privacy awareness training for all customer-facing staff.
- Scheduled annual external compliance audit.
- Published a public privacy transparency report summarizing data usage and consumer rights.

Cross-Functional Implementation: Governance Accountability Matrix

Privacy governance necessitates coordination across all functions that collect, process, or make decisions based on customer data. Unlike traditional compliance approaches that isolate legal responsibilities, systematic governance embeds privacy considerations throughout all business operations. When privacy requirements evolve or consent patterns change, coordinated teams must adjust policies, update systems, and maintain customer trust simultaneously. The following accountability matrix ensures that governance becomes an organizational capability rather than a responsibility of the legal department.

- **Data Steward Council:** Owns overall governance strategy, policy development, and cross-functional compliance coordination. Maintains regulatory monitoring and strategic privacy decision authority.
- **Legal/Compliance:** Develops privacy policies, manages regulatory interpretation, and oversees audit processes. Provides legal guidance and ensures regulatory alignment across all departments.
- **Marketing Operations:** Implements consent management systems, maintains preference centers, and ensures campaign compliance. Monitors engagement patterns for trust indicators and consent quality.

- **IT Security:** Manages technical privacy controls, data encryption, and security incident response. Provides infrastructure protection and breach prevention systems.
- **Training/HR:** Coordinates privacy education, compliance training, and staff certification programs. Ensures organizational privacy competency and awareness maintenance.

Promolytics in Practice: Governance, Compliance & Trust

What We Can Do Today:

- **Consent Capture at Source** – Every QR code or link survey includes explicit consent language and the consumer opts-in before data is collected.
- **Secure Storage** – Customer data access is role-restricted in our database.
- **Privacy Policy Transparency** – Public documentation on what data is collected and how it's used (available for anyone to read, so customers know how their information is handled).

What's Coming Next:

- **Automated Opt-Out Propagation** – Opt-out requests will be pushed automatically across all linked systems within 24 hours.
- **Real-Time Privacy Dashboard** – Clients will be able to monitor consent rates, opt-out trends, and compliance status from their dashboards.

- **Dynamic Preference Portals** – Consumers will be able to adjust their data-sharing preferences (that is, what personal information they permit to be collected and used) in real-time.
- **Quarterly Compliance Alerts** – Automated reminders for clients when state or federal laws change that may impact data collection workflows.

Why It Matters:

Even the most actionable marketing insight can erode brand trust if it's collected or managed improperly. By embedding governance into the same workflow as campaign tracking and analysis, Promolytics ensures that insight generation and trust-building happen in lockstep, not as competing priorities.

With these guardrails in place, Northbridge isn't just building smarter campaigns; it's compounding trust capital, ensuring that every data-driven win can be sustained and scaled.

Executive Reflection – Trust Dividend:

Which single governance metric, if missed, would most likely damage customer trust, and what preventive control can be implemented to ensure adherence to it?

Chapter 20 Checklist: Embedding Governance and Trust

- Update consent language at all data capture points for clarity.
- Verify opt-out propagation across every downstream system within 24 hours.
- Conduct a data mapping exercise to confirm no orphan stores remain.

- Schedule an external penetration test and secure a budget.
- Publish an annual privacy summary to reaffirm trust.

With governance embedded into every stage of the INSIGHT loop, Northbridge can now scale its data strategy with confidence, knowing that every campaign, every model, and every decision is backed by a foundation strong enough to withstand both regulatory scrutiny and customer expectations. This trust infrastructure becomes the platform for advanced competitive positioning and market leadership that distinguishes truly data-driven organizations from those that merely collect information.

Chapter Twenty-One

Owning the Insight:

When Data Mastery Transforms Into Market Leadership

Organizational transformation from data collection to competitive intelligence represents the defining business capability of the next decade. Companies that master systematic first-party data strategies don't just improve marketing efficiency—they create sustainable competitive moats through customer insight ownership, predictive market positioning, and trust-based relationship building.

This book provides a comprehensive operational framework for building insight-driven competitive advantage, spanning foundational data capture through advanced AI capabilities, from departmental execution to executive leadership, and from compliance management to customer trust development. The INSIGHT framework, maturity model, and departmental playbooks create systematic capability advancement rather than project-based improvements.

Success requires sustained organizational commitment to the habits, processes, and cultural shifts that maintain competitive advantage. Companies that implement these frameworks systematically while protecting against capability drift will compound their market position with every cycle, creating increasingly difficult competitive moats for rivals to overcome.

Northbridge: The Complete Transformation

Two years later, the Northbridge boardroom tells a different story. Where disconnected departmental reports once dominated quarterly reviews, a unified dashboard now displays real-time customer intelligence that drives every strategic decision. The transformation metrics speak for themselves:

Revenue Impact: 34% increase in customer lifetime value, 28% improvement in new product success rates, 41% reduction in customer acquisition costs compared to pre-transformation baselines.

Operational Excellence: Marketing campaigns achieve 5.2x higher incremental lift, sales cycles are shortened by 60%, and product launches reach profitability 40% faster than industry averages.

Competitive Positioning: The signal-to-strategy lag has been reduced from 120 days to 30 days, enabling market moves that competitors can't match. Three major product innovations were launched before competitors identified the opportunities.

But the numbers only tell part of the story. The deeper transformation lies in the organization's culture and capabilities.

"We no longer just have siloed marketing meetings, sales meetings, and product meetings," Evelyn reflects. "We also have intelligence meetings where every department contributes to and acts on the same customer insights. The Flywheel isn't just spinning; it's accelerating."

The cultural shift became visible in daily operations. Marketing campaigns launch with sales field intelligence and product roadmap alignment. Sales presentations incorporate real-time local market data that marketing teams have captured. Product development priorities reflect validated consumer needs that sales confirmed in buyer conversations.

Most importantly, Northbridge's customer relationships deepened rather than becoming more transactional. Consent rates increased to 91%, trust metrics showed consistent positive trends, and voluntary customer engagement grew 67% as transparent value exchange replaced hidden data extraction.

The transformation required 18 months of disciplined execution, $2.3M in infrastructure investment, and organizational change management that impacted every role. However, the result created competitive advantages that compound with each quarter: better customer intelligence leads to more effective campaigns, which generate higher trust, enabling deeper insights, which in turn fuel superior product development.

"The hardest part wasn't building the systems," notes Jasmine Lee. "It was maintaining the discipline to use them consistently when quarterly pressures tempted shortcuts. But every time we protected the process, the results got stronger."

Today, Northbridge competitors struggle to match their market timing, product relevance, and customer loyalty because they lack the systematic insight generation that Northbridge has built. The competitive moat isn't just data—it's the organizational capability to turn customer signals into market advantage faster and more accurately than rivals.

Implementation Sustainability Framework

Sustaining insight-driven competitive advantage requires systematic protection against capability drift and cultural regression. Organizations must institutionalize the habits and governance that maintain advancement:

Leadership Accountability Protocols

- **Monthly Executive Reviews:** INSIGHT framework performance across all departments, focusing on process adherence and outcome quality rather than just results

- **Quarterly Maturity Assessments:** Systematic evaluation using Chapter 17's framework to prevent regression and identify advancement opportunities

- **Annual Strategic Realignment:** Complete review of competitive positioning, customer trust metrics, and organizational capability development

Organizational Change Protection

- **Role-Based Accountability:** Every position includes specific INSIGHT framework responsibilities in job descriptions and performance reviews
- **Cross-Functional Integration Metrics:** Measurement systems that reward collaboration and shared data sources rather than departmental optimization
- **Cultural Reinforcement Programs:** Regular training, success story sharing, and process improvement initiatives that strengthen insight-driven decision making

Capability Maintenance Systems

- **Technology Infrastructure Oversight:** Continuous monitoring of data quality, system integration, and governance compliance to prevent technical debt
- **Process Audit Cycles:** Regular review of departmental playbooks, escalation protocols, and decision frameworks to ensure consistent execution
- **Competitive Intelligence Updates:** Systematic monitoring of market changes, regulatory developments, and competitive responses that re-

quire framework adjustments

Cross-Functional Implementation: Transformation Sustainability Matrix

Long-term competitive advantage requires every function to maintain and advance insight-driven capabilities rather than reverting to previous practices. Sustainable transformation embeds accountability for maintaining capability throughout the organization, while protecting against the natural tendency toward fragmentation and shortcuts. The following accountability matrix ensures transformation sustainability becomes organizational muscle memory rather than a temporary process improvement.

- **Executive Leadership:** Owns strategic vision maintenance, resource allocation for capability advancement, and organizational culture protection. Maintains long-term competitive positioning focus.

- **Data Governance Council:** Ensures adherence to a systematic framework, process improvement, and capability development. Prevents regression through continuous monitoring and corrective action.

- **Department Champions:** Lead functional implementation of the INSIGHT framework, playbook execution, and cross-departmental coordination. Serve as change agents and process advocates.

- **Training and Development:** Coordinates ongoing capability development, new employee onboarding, and organizational learning. Ensures systematic knowledge transfer and skill advancement.

- **Performance Management:** Integrates insight-driven metrics into individual and team evaluation systems. Rewards behaviors that strengthen rather than undermine systematic capabilities.

Final Implementation Roadmap

First 90 Days: Foundation Protection

- Conduct a comprehensive post-implementation audit of all INSIGHT framework components
- Identify and address any process gaps or capability weaknesses before they compound
- Establish sustainability metrics and monitoring systems for long-term capability maintenance
- Document lessons learned and create organizational playbook for capability protection

Months 4-12: Capability Advancement

- Progress one full maturity stage across core business functions using Chapter 17's advancement framework
- Implement advanced AI capabilities and predictive analytics where foundational infrastructure supports advancement
- Expand Consumer Insight Collection Strategy to additional customer touchpoints and market segments
- Develop competitive intelligence systems that monitor market changes and competitive responses

Year 2 and Beyond: Competitive Moat Expansion

- Achieve "Intelligent" maturity stage in core business functions, enabling real-time optimization and predictive market positioning
- Scale successful frameworks to additional product lines, market segments, or geographic regions
- Develop proprietary analytics capabilities that create sustainable competitive advantages
- Build organizational learning systems that continuously improve insight generation and application effectiveness

Executive Reflection – Transformation Legacy

Looking ahead twelve months, which single capability—systematic insight generation, cross-functional coordination, or competitive intelligence—represents your organization's greatest opportunity for sustainable advantage, and what specific investment will you make to ensure that capability compounds rather than stagnates?

Final Implementation Checklist: Sustaining Competitive Advantage

- Establish monthly executive reviews focused on the INSIGHT framework performance and process adherence.
- Integrate insight-driven responsibilities into every job description and performance evaluation system.
- Create capability maintenance budgets that protect infrastructure investment and enable continuous advancement.

- Document a complete transformation playbook for organizational learning and knowledge transfer.
- Establish competitive intelligence monitoring systems that identify market changes requiring framework adjustments.
- Schedule annual strategic reviews that realign capabilities with the evolving competitive landscape.
- Build cross-functional leadership development programs that strengthen insight-driven decision making.
- Create customer trust monitoring systems that ensure sustainable relationship development.
- Establish vendor and technology partnership criteria that support rather than undermine systematic capabilities.
- Develop succession planning that preserves an insight-driven culture through leadership transitions.

The companies that master this systematic approach don't just implement better analytics; they build organizational capabilities that compound competitive advantage with every cycle. Customer insights become deeper, market timing becomes more precise, and competitive responses become faster and more accurate.

Northbridge's transformation proves that first-party data excellence isn't about technology or tactics; it's about systematic organizational capability that turns customer intelligence into sustainable market leadership. The framework works, but only for organizations committed to the disciplined execution that makes competitive advantage sustainable rather than temporary.

Insight ownership creates a lasting competitive advantage, but only when protected by the systems, culture, and leadership commitment that prevent regression and enable continuous advancement. Master the framework, protect the culture, and maintain the discipline. The competitive moat will compound with every cycle.

Appendices

Scan the QR code to receive a pdf download of the Appendices.

Appendix A · Overall Evaluation Criterion (OEC) Selection Worksheet

Overall Evaluation Criterion (OEC) Definition. A single primary metric—often a weighted composite—that quantifies the objective of a test and serves as the deciding score for winners. A good OEC is measurable in the short term yet predictive of long-term business value, so you can experiment quickly without optimizing for vanity signals.

What makes a good OEC

- **Sensitive** to changes that matter
- **Timely** within the experiment window
- **Attributable** per variant/region/cohort
- **Relevant** to durable outcomes (profit, retention, LTV)

OEC vs. guardrail metrics

The **OEC** decides the winner. **Guardrails** are "safety" metrics that you monitor to prevent collateral damage, such as page speed, unsubscribe rate, or customer support contacts. You can ship only if the OEC improves and guardrails remain within limits.

OEC Selection Worksheet

Use this worksheet before every campaign or experiment to define the Overall Evaluation Criterion (OEC) and guardrails. Keep it to one page.

1) Business objective & context

Objective (revenue, retention, trial, etc.)	
Target segment / region	
Offer / incentive	
Expected time-to-impact (days)	
Channels (on/offline) to measure	
Data sources for verification (POS, receipts, CRM, etc.)	

2) Candidate OECs (list up to 3 and compare)

Metric (OEC candidate)	How verified	Time window	Pros	Risks / bias

3) Guardrail metrics (monitor but do not optimize for)

Metric	Limit / threshold	Why it matters

4) Decision rule

Winner if OEC ≥ __________ compared to control/holdout AND all guardrails within limits for the full analysis window.

5) Experiment design essentials

Design: ▒ A/B holdout ▒ Geo test ▒ Synthetic control ▒ Other: __________
Minimum sample size / power: __________________________
Analysis window (days): __________________________
Pre-registered before launch: ▒ Yes ▒ No

6) Composite OEC (optional example)

OEC = w1·(incremental verified purchases) + w2·(repeat within 60 days) + w3·(margin per unit) – w4·(promo cost per verified buyer) – w5·(OOS penalty)

7) Approvals

Owner (Marketing)	Owner (Finance)	Date

Note: Choose an OEC that is measurable in the short term yet predictive of long-term value. Guardrails prevent wins that harm the business.

Appendix B · Calculate Your Promotion Blind-Spot Worksheet

Use these worksheets to quantify the "blind spots" in your promotional campaigns – areas where ROI isn't measured or customer data isn't captured. By filling out these worksheets, marketing managers and brand strategists can identify how much of their promotion effort is going untracked and uncover missed data opportunities. This will help highlight the ***percentage of promotions with unmeasured ROI, missed first-party data opportunities****, and the* ***cost of untracked campaigns****.*

Note: A lack of measurement is a widespread issue; marketing ROI measurement remains a persistent challenge across the industry. Also, failing to capture customer data is a lost opportunity, especially since Most marketing professionals now view first-party data collection as essential to their strategy. Keep these points in mind as you assess your promotion blind spots.

Worksheet for Marketing Managers

Designed for marketing managers responsible for campaign budgets and ROI. This section focuses on promotional spend, sales lift, and ROI tracking.

1. **Document Your Promotion Totals:** Start by gathering high-level figures for your recent promotions (e.g., those from the past year or quarter). Fill in the following:

    - **Total Promotional Budget:** $_____ (the overall budget allocated for promotions in the period).

    - **Actual Promotional Spend:** $_____ (the amount actually spent on all promotions; use the same period as the budget).

    - **Total Sales Lift from Promotions:** $_____ (the total incremental sales generated by promotions where you *did* measure results).

These figures set the stage. "Sales Lift" should reflect the added revenue from promotions that had measurable outcomes. If some promotions weren't tracked, their sales impact won't be in this number – highlighting part of your blind spot.

1. **Assess ROI Tracking for Each Promotion:** List the individual promotions and indicate whether their ROI was measured or not. For each campaign/promotion in the period, note:

    - **Promotion Name or Description – Spend $ – ROI Measured?** (Yes/No) – **Sales Lift** (if measured) – **First-Party Data Collected?** (Yes/No).

For example, you might have: *"Spring Sale – $50,000 spend – ROI Measured: Yes – $120,000 sales lift – Data Collected: Yes (emails gathered)"* or *"Holiday Campaign – $30,000 spend – ROI Measured: No – Sales lift: Unknown – Data Collected: No."* Go through all your promotions in this manner.

- **Count the promotions with measured ROI:** ____ campaigns had

Yes for ROI Measured.

- **Count the promotions with unmeasured ROI:** ____ campaigns had **No** for ROI Measured.

Now, calculate the **percentage of promotions with unmeasured ROI**. Use the counts above:

Unmeasured ROI % = (Number of promotions with no ROI measurement ÷ Total number of promotions) × 100

Fill in your result: **Unmeasured ROI Promotions:** ____ out of ____ total campaigns = _____**%** without measured ROI.

A high percentage of unmeasured ROI means much of your campaign's effectiveness remains unknown.

1. **Calculate the Cost of Untracked Campaigns:** Next, quantify how much budget was spent on promotions that **lacked real-time ROI tracking**. Using your list from step 2, add up the spend for all campaigns where real-time ROI was **not** measured:

    - **Total Spend on Untracked Promotions:** $_____ (sum of budgets for all "ROI not measured" campaigns).

This dollar amount represents the **cost of untracked campaigns**, which is money spent without clear real-time ROI accountability. Write down what portion of your overall budget this untracked spend represents: **Untracked Spend as % of Budget** = (Untracked Spend ÷ Total Promotional Budget) × 100 = ____%.

- **Known Sales Lift from Tracked Promotions:** $_____ (sum of sales lift from the campaigns where ROI was measured, from step 2). *(Optional: Calculate the ROI for the tracked portion – e.g., known sales lift ÷ spend on tracked campaigns – to see how effective your measured campaigns were. This gives an idea of what the untracked portion might be missing.)*

These are **missed first-party data opportunities** – campaigns where you could have gained customer contacts or insights but didn't. Jot down any notable details (e.g., "Spring Sale had 500 shoppers but no emails captured"). While a brand strategist will delve deeper into this, as a manager, you should recognize how many campaigns lacked a data component. (Given the importance of first-party data in marketing today, each missed opportunity can hinder future targeting and personalization.)

1. **Summarize Your Promotion Blind-Spot (Manager View):** In the section below, fill in the key findings from your analysis:

   - **% of Promotions with Unmeasured ROI:** ____% (from step 2 calculation).

   - **Cost of Untracked Campaigns:** $_____ (and ____% of budget) was spent on promotions without ROI tracking.

   - **Missed First-Party Data Opportunities:** ____ promotions (where no customer data was captured).

Example summary: "40% of our promotions had no ROI measured, accounting for $200,000 of spend with unknown returns. We also found that 5 out of 12 campaigns collected no customer data, representing significant missed data opportunities."

Take a moment to reflect on these numbers. A high unmeasured ROI percentage or significant untracked spend indicates a **promotion blind spot** – you're effectively "flying blind" for a substantial share of your marketing budget. Likewise, each campaign that fails to collect customer information is a blind spot in building your customer database. These insights can help you justify improvements (e.g., investing in better tracking tools or requiring data capture in every promotion).

Worksheet for Brand Strategists

Designed for brand strategists focusing on customer engagement and data. This section highlights the capture of first-party data, the promotional impact on the brand/customer base, and the extent to which campaign performance is not accurately measured.

1. **Inventory Your Promotions and Data Capture:** List all recent promotions and note their audience engagement elements. For each campaign, record:

    ◦ **Promotion Name – Objective/Offer – Customer Data Collected?** (Yes/No and what type, e.g., emails, sign-ups) – **Outcome Tracked?** (Yes/No for ROI or any success metric).

For example: *"Spring Sale – 20% off – Data Collected: Yes (signup form for coupon, 300 emails) – Outcome Tracked: Yes (sales uplift measured)"*, or *"Brand Launch Event – new product intro – Data Collected: No – Outcome Tracked: No formal tracking beyond attendance."* Include all promotions that occurred during the period under review. This provides a clear picture of which campaigns served as data-collection opportunities and which did not.

- **Count of promotions that collected first-party data:** ____ campaigns (those marked Yes for data collected).

- **Count of promotions that did not collect data:** ____ campaigns.

These "No data" campaigns are your **missed first-party data opportunities**. For each of these, consider what was lost – e.g., *"no email leads from 500 participants"* or *"no customer insights gathered during a major giveaway."* This highlights how many opportunities to enrich your customer database or gain valuable insights were missed.

1. **Evaluate ROI/Performance Tracking:** Now, identify how many promotions had **no performance measurement** (no ROI or sales lift

tracked, no clear KPI outcomes). This is important for a brand strategist because without tracking, you lose insight into what resonates with your audience and what drives brand value. Using your inventory:

- **Promotions with ROI or outcome measured:** ____ campaigns (those marked Yes for outcome tracked – e.g. you have sales data, engagement metrics, or a clear success indicator).
- **Promotions with no measurement of results:** ____ campaigns.

Calculate the **percentage of promotions with unmeasured ROI** (or unmeasured outcomes):

Unmeasured Outcome % = (Promotions with no performance tracking ÷ Total promotions) × 100

Record your result: **Untracked Promotions:** ____ out of ____ = ____**%** with no measured outcome.

This is a strategic blind spot – if, for example, 50% of campaigns weren't measured, that means half of your promotional initiatives have no data on whether they boosted sales, awareness, or customer engagement. *It's hard to refine brand strategy when results aren't tracked.*

1. **Quantify the Impact of Untracked Campaigns:** Even if exact revenue isn't the primary focus of a brand strategist, it helps to know the scale of resources put into unmeasured efforts. Work with your marketing team or estimates to determine:

   - **Estimated Spend on Untracked Campaigns:** $_____ (the total budget spent on the promotions that had no outcome tracking).

This figure tells you the **cost of campaigns run without insight**. In other words, $X was spent without firm knowledge of the return or brand impact. You can also note this as a percentage of the total promotion budget: **Untracked Spend as % of Budget:** ____%. A high number here signals risk, marketing dollars spent without learning what works. (Marketing studies have found that poor tracking can lead to wasted spend.)

- **Qualitative Brand Impact:** For each untracked campaign, briefly consider the potential brand effects you **could** have measured. For example, *"Campaign X (untracked) aimed to boost brand awareness, we didn't measure it, so we don't know if awareness grew."* Note these missed insights. While not a number, it's essential to recognize what strategic learnings were lost along with the data.

Example summary: "We found that 4 of 10 promotions (40%) collected no customer data – missed opportunities to build our CRM. Additionally, 50% of campaigns weren't tracked for success metrics, representing about $150,000 of spend with no clear ROI. These blind spots leave us unsure which promotions truly drove engagement or sales."

Missed data means fewer insights for personalization and retention. Untracked outcomes mean you don't know which campaigns strengthen brand relationships or perception. Address these by ensuring every promotion has a data capture element and a defined success metric. Use this worksheet to quantify your blind spots and plan remediation.

Appendix C · Data Source Comparison Guide

First-Party Data:

- **Source**: Collected directly by your organization from your customers or users
- **Examples**: Website analytics, purchase history, email subscribers, loyalty program data, survey responses, and customer service interactions
- **Ownership**: You own and control this data
- **Quality**: Higher quality and accuracy since it comes from direct interactions
- **Cost**: Typically, a lower cost to collect
- **Privacy**: Fewer privacy concerns since customers are knowingly sharing it with you
- **Targeting**: Highly relevant for your specific business and customer base

Second-Party Data:

This is essentially another company's first-party data that they share directly with you through partnerships, rather than through a data marketplace.

Third-Party Data:

- **Source**: Collected by external organizations and purchased or accessed through them

- **Examples**: Demographic data from data brokers, social media insights, market research datasets, and advertising network data
- **Ownership**: You license or purchase access, but don't own the underlying data
- **Quality**: Can vary significantly; may be less accurate or outdated
- **Cost**: Usually involves licensing fees or purchase costs
- **Privacy**: More complex privacy considerations and regulatory restrictions
- **Scale**: Often provides a broader reach and additional context

The trend in recent years has been toward prioritizing first-party data strategies due to privacy regulations, browser changes that may limit third-party cookies, and the higher quality and compliance benefits of data collected directly from customer relationships.

Appendix D · Recommended Tools & Resources

Category	Example Tools	Purpose
Capture Data & User Experiences	Promolytics (promolytics.net), Typeform	Collect first-party data via QR, URL, forms
Identity Resolution	Customer Data Platforms (CDPs), Promolytics (promolytics.net)	De-duplicate and unify profiles
Analytics & BI	Looker, Tableau, Promolytics (promolytics.net)	Visualize and analyze unified data
Activation	ESPs, CRM Automation, Promolytics (promolytics.net)	Deliver personalized journeys
Governance	Consent management platforms, Promolytics (promolytics.net)	Manage permissions, DSAR workflows
Security	Encryption & monitoring tools	Protect data at rest and in transit

Appendix E · QR Code Statistics & Placement Playbook

Statistics

- Marketer usage momentum: 93% of marketers increased QR use in the past 12 months, 86% plan to increase further.[1]
- In-store usage: 64% of shoppers have scanned a QR code while shopping in-store.[2]
- "~100M US smartphone users scanning QR codes by 2025."[3]
- Sunrise 2027 is the global industry commitment, led by GS1, to transition retail and healthcare point-of-sale/point-of-care systems to accept

---

1. Bitly "From Scans to Strategy: How Marketers Use QR Codes in 2025" marketer survey.
2. 1WorldSync Product Content Benchmark blog summary. Consumer Survey Data: The Growing Impact of QR Codes and AI on In-Store Shopping Experiences December 10, 2024
3. Usability of Quick Response (QR) Code as a Method to Access a Survey Using a Smartphone. Center for Behavioral Science Methods Research and Methodology Directorate U.S. Census Bureau Washington, D.C. 20233 August 2024
4. GS1: https://www.gs1ie.org/blog/2024/joint-industry-statement-qr-codes-powered-by-gs1-standards.html Joint Industry Statement: QR codes powered by GS1 standards. June 26, 2024

2D barcodes—specifically QR codes powered by GS1 standards (GS1 Digital Link)—by the end of 2027. The shift is already well underway: pilots are active in 48 countries, covering approximately 88% of global GDP, while brands begin printing 2D codes on-pack and retailers upgrade scanners to read both traditional 1D barcodes and 2D codes during the transition. For marketers, this unlocks a single on-pack gateway for consumer engagement, coupons, product details, and recalls; for operations, it enables richer data (GTIN plus application identifiers like lot and expiration) to flow through supply chains and POS.[4]

Placement Playbook

Here's a comprehensive, CPG-ready checklist of places to put QR codes on marketing materials. Grouped by channel and production workflow.

On-product and packaging:

- Primary label or sleeve
- Neck hanger or hang tag
- Cap, lid, or tamper seal
- Secondary packaging: cartons, multipacks, shrink wraps
- Box interior flap or tray
- Peel-to-reveal sticker for promos or compliance copy

---

4. GS1: https://www.gs1ie.org/blog/2024/joint-industry-statement-qr-codes-powered-by-gs1-standards.html Joint Industry Statement: QR codes powered by GS1 standards. June 26, 2024

- Product inserts and instruction leaflets

Retail point-of-sale (in-store):

- Shelf talkers and wobblers
- Price tags and promo tags
- Aisle violators and category blades
- Endcap signage and header cards
- Cooler, freezer, or display case clings
- Floor decals near the product
- Checkout counter toppers and queue stanchions
- Self-checkout screens and kiosk decals

On-premise and hospitality:

- Menus and drink lists
- Table tents and check presenters
- Coasters, placemats, and napkin bands
- Tap handles or dispenser decals
- Bar mats and back-bar signage
- Host stand and entrance signage
- Room drops and minibars (for hotels)

Events, demos, and sponsorships:

- Demo table signs and tasting mats
- Sampling cups or sleeves
- Staff badges, lanyards, and wristbands
- Photo booth backdrops and props
- Step-and-repeat banners
- Event programs and schedules
- Pop-up banners, feather flags, and tents
- Giveaways, swag bags, and insert cards

Out-of-home and transit:

- Posters, wild postings, and window clings
- Bus shelters and transit station posters
- Street kiosks and digital OOH overlays
- Mall signage and elevator wraps
- Gas-pump toppers and forecourt screens
- Ride-share car toppers and airport displays
- Building lobby screens and directory boards

Direct mail, print, and collateral:

- Postcards and self-mailers
- Catalog covers and inside front covers
- Brochures, one-pagers, and spec sheets
- Magazine and newspaper ads
- Door hangers and take-one flyers
- Business cards and appointment cards

E-commerce, post-purchase, and fulfillment:

- Packing slips and invoices
- Shipping labels and box exteriors
- Box interior panels for unboxing moments
- Return label sheets and instruction cards
- Thank-you cards and loyalty inserts

Receipts, POS systems, and payments:

- Printed receipt header or footer
- Digital receipt emails and SMS
- Payment terminal welcome or idle screens
- Kiosk screens and order confirmation slips

Partners and co-marketing:

- Delivery bags and pizza boxes
- Coffee sleeves and bakery bags
- Meal kits and subscription boxes
- Community boards and partner windows
- Charity or cause-marketing materials

Vehicles, equipment, and environments:

- Delivery vans and service vehicles
- Refrigerators, vending machines, and kiosks
- Trade-show booths and counters
- Office or storefront windows and doors
- Restroom mirrors and stall posters
- Stair risers and escalator handrail covers

Digital cross-overs (when printed or DOOH is present):

- CTV and livestream lower-thirds during sponsored segments
- In-venue jumbotrons and arena ribbons
- YouTube or webinar interstitials, when displayed on a large screen

- TV advertising

Quick placement tips:

- Minimum size for arm's-length scanning: about 0.8–1.2 inches square for print with a clear, quiet zone.
- High contrast is critical. Dark code on a light background is most effective.
- Add a short, specific call to action and a fallback short URL printed below.
- Use dynamic QR codes with tracking parameters to attribute scans by location, partner, and creative.
- Match the landing page to the context. A tasting table should land on a fast poll or offer, not a generic homepage.
- Include consent language where needed and age-gate when applicable.

Appendix F · INSIGHT Agenda Template

Cadence Tier	Timing	INSIGHT Flow	Purpose & Key Outputs
1. Individual Campaign Review	First work week after the campaign has finished	1. Identify & normalize 2. Segment 3. Interpret 4. Guide decisions 5. Harness automations 6. Track	• Verify data capture is clean and complete. • Update audience segments with signals. • Surface one clear insight/storyline. • Approve optimizations and queue automations. • Circulate a quick-hit report; note any learning velocity issues.
2. Portfolio Review – All Active Campaigns (Quarterly)	One week after each fiscal quarter closes	Same six-step INSIGHT loop applied across the entire active-campaign set	• Spot cross-campaign anomalies and migration trends. • Publish a consolidated "Quarter in Review" deck. • Re-score segment health and growth. • Prioritize mid-quarter tests/budget shifts. • Refresh automation rules that span multiple campaigns.
3. Year-End & YOY Review – All Active Campaigns	Early January (or immediately after the fiscal year close)	INSIGHT loop & Year-over-Year comparison for recurring campaigns	• Compare key KPIs YOY (lift, ROI, redemption, LTV). • Highlight evergreen vs seasonal performance patterns. • Lock in strategic changes for the next annual planning cycle. • Archive data snapshots and audit compliance. • Issue an executive summary with a recommended 12-month roadmap.

INSIGHT Timeline

Appendix G · US Privacy Timeline & State Matrix

US State Privacy Timeline (2018–2025)

Year	Event	Key Impact on Marketers
2018	GDPR (EU) takes effect	Sets global tone for consent and data-subject rights
2020	California Consumer Privacy Act (CCPA) takes effect	Right to know, delete, and opt out of sale/share
2021	Virginia (VCDPA) & Colorado (CPA) enact comprehensive statutes	Broaden scope beyond California
2022	California Privacy Rights Act (CPRA) amends CCPA	Stricter enforcement; new sensitive data category; right to correct
2023	Colorado & Connecticut go into effect; Utah goes into effect (Dec 31)	Patchwork regime arrives; multi-state consent/opt-out management needed
2024	Texas (TDPSA) & Oregon (OCPA) go into effect (Jul 1); Montana (MTCDPA) goes into effect (Oct 1)	Retail/e-commerce teams face broader US coverage; prepare for GPC/UOOM signals where applicable
2025	Jan 1: Delaware (DPDPA), Iowa (ICDPA), Nebraska (NDPA), New Hampshire (SB255) effective; Jan 15: New Jersey (NJDPA) effective	Expand consent/opt-out flows to new states; align data maps and notices
2025	Jul 1: Tennessee (TIPA) effective; Jul 31: Minnesota (MCDPA) effective; Oct 1: Maryland (MODPA) effective	Strengthen DPIAs, sensitive-data handling, and opt-out signal handling; update retention & minimization policies
2025 (also)	CO: Biometric obligations (Jul 1) & minors protections (Oct 1); OR: 501(c)(3) coverage begins (Jul 1)	Ensure UOOM/GPC handling and age/minor controls where scope applies
Source: IAPP US State Privacy Legislation Tracker; state AG pages and law firm summaries.		

2025 State Privacy Matrix (Newly Effective in 2025)

State	Law (short name)	Effective date	What teams must ensure (quick take)
Delaware	Delaware Personal Data Privacy Act (DPDPA)	1-Jan-25	Honor access/correct/delete + opt-out of sale/targeted ads; DPIAs by 7/1/2025; honor universal opt-out signals starting 1/1/2026
Iowa	Iowa Consumer Data Protection Act (ICDPA)	1-Jan-25	Baseline rights (access/correct/delete/opt-out); update notices; 90-day cure window applies
Nebraska	Nebraska Data Privacy Act (NDPA)	1-Jan-25	Modeled on Texas; rights to access/correct/delete/opt-out; update notices and consent routing
New Hampshire	SB 255 (Expectation of Privacy Act)	1-Jan-25	Access/correct/delete + opt-out of sale/targeted ads; recognize opt-out signals beginning 1/1/2025
New Jersey	New Jersey Data Privacy Act (NJDPA)	15-Jan-25	Access/correct/delete/port + opt-out; proposed rules published 6/2/2025—watch rulemaking for details
Tennessee	Tennessee Information Protection Act (TIPA)	1-Jul-25	Virginia-style framework; NIST privacy-framework safe harbor; ensure opt-out links/appeals
Minnesota	Minnesota Consumer Data Privacy Act (MCDPA)	31-Jul-25	Strong data minimization and purpose-limitation emphasis; detailed policy & DPIA expectations
Maryland	Maryland Online Data Privacy Act (MODPA)	1-Oct-25	Strict data minimization and sensitive-data limitations; recognize opt-out preference signals on 10/1/2025
Focus: States that first become effective during 2025 (plus unique notes that affect marketing, CRM/CDP, and consent flows).			

References

- IAPP - US State Privacy Legislation Tracker(updated July 7, 2025):https://iapp.org/resources/article/us-state-privacy-legislation-tracker/
- IAPP – Key Dates 2025 (CO biometrics &minors; OR nonprofit coverage): https://iapp.org/resources/article/iapp-key-dates/
- Colorado AG - UOOM page: https://coag.gov/uoom/

- Texas AG - TDPSA effective July 1, 2024: https://www.texasattorneygeneral.gov/consumer-protection/file-consumer-complaint/consumer-privacy-rights/texas-data-privacy-and-security-act
- Oregon DOJ - OCPA effective July 1, 2024 (nonprofits July 1, 2025): https://www.doj.state.or.us/consumer-protection/id-theft-data-breaches/privacy/
- IAPP - State detail pages: Delaware, Iowa, Nebraska, New Hampshire, New Jersey, Tennessee, Minnesota, Maryland (see state 'Key dates' sections).

Appendix H · Marketing → Sales Handoff Template

Marketing-to-Sales Handoff Kit

Weekly cadence

Monday: Marketing completes analysis and publishes handoff kit, tags accounts in the CRM, and prepares for a sync to align talking points, objections, & next best offers.
Tuesday 2:00 p.m.: Marketing-Sales alignment meeting (30 minutes).
Wednesday-Thursday meeting: Sales reps customize materials for specific accounts.
Customer Meeting: Lead with local evidence and a specific expansion ask.
Post-Customer Meeting: Log Execution Notes and Update Risk Flags.

Deck structure

Slides 2–3: Store Heatmap (Buyer Brief + CSV Schema)
Slide 4: Account Brief
Slide 5: Lift Card (one per activation)
Slide 6: Execution Notes (field checklist)
Slide 7: Risk Flags Register
Slide 8: KPI & Formula Reference

Store Heatmap - Buyer Brief

Retailer-facing snapshot for local performance

Fill these fields

1. Retailer Name: __________
2. Date Range: __________
3. Top Stores by Incremental Lift: 1) ___ 2) ___ 3) ___
4. Best-performing Activation: __________
5. Winning Message Theme: __________
6. Conversion Path: Scan -> Redemption: ___%
7. Display Compliance Average: ___%
8. Stockout Incidents: ___
9. Local Adjustments: __________
10. Suggested Actions Today: SKUs to add, display count, next promo window

Drop Heatmap Screenshot Here

Store Heatmap - CSV Schema

Use this exact column order for joins and deck automation.

column_name	type	example	notes
retailer_id	text	RLT-1032	Unique per chain or independent
retailer_name	text	Eastside Market	
store_id	text	ES-PA-014	
store_name	text	Eastside Market South	
address	text	123 Main St	
city	text	Philadelphia	
state	text	PA	2-letter
zip	text	19103	
latitude	number	39.9526	
longitude	number	-75.1652	
campaign_id	text	AUTUMN-BUNDLE-24	

Account Brief - Template

Taylor per retailer before the meeting.

Account Header

- Retailer: __________
- Meeting Date: __________
- Owner: __________
- Quarter Objective: __________

Shopper Insights

- Top Segments: __________
- Preferences: __________
- Repeat Interval: __________
- Channel Response: __________

Plan & Proof

- Assortment & Promotions: __________
- Displays & Field Focus: __________
- Verified Lift Last Cycle: ___% (CPIU: $__)
- Proof Links: charts, photos, test design

The Ask

- Approve SKU adds: ______________
- Approve display count: ______________
- Next promo window: ___________

Anticipated objections -> responses:
1) ______________ -> ______________
2) ______________ -> ______________

Lift Card - Template

One card per activation

Experiment Summary

Activation: __________ | Dates: __________
Channel: __________ | Audience: __________
KPI: __________
Hypothesis: If __________ then __________ by ___%
Design: Treatment N=___, Control N=___,
Randomization: __________, Holdout: ___

Results

Treatment conv: ___% | Control conv: ___%
Incremental lift: ___%
Incremental units: ___ | CPIU: $__ | ROI: ___
Confidence/stat result: __________
Bias checks: __________
Decision: Scale / Iterate / Retire
Next change: __________

Drop Control vs. Treatment Chart Here

Execution Notes - Field Checklist

Complete after each visit.

Field	Value / Notes
Visit Date	__________
Account & Store	__________
Rep	__________
Campaign	__________
Placement correct (Y/N)	__________
Count meets plan (Y/N)	__________
Materials correct (Y/N)	__________
Photos (links)	__________
Price matches plan (Y/N)	__________
Promo price live (Y/N)	__________
Facings (#)	__________
Stockouts (Y/N)	__________

Risk Flags Register - Template

Track issues that threaten performance.

risk_id	type	trigger condition	severity	impact	evidence	mitigation	owner	SLA_ days	next_che ck
RF-001	Data quality	Missing receipt validation > 8%	High	Cannot verify lift	QA report link	Reprocess & tighten validation	Owner	2	YYYY-MM-DD
RF-002	Execution	Compliance score < 80 in top stores	Medium	Understates potential	Photo set link	Merch sweep in 72h	Owner	3	YYYY-MM-DD
RF-003	Inventory	Stockout flag > 10% stores	High	Lost sales	POS feed link	Expedite replenishment	Owner	1	YYYY-MM-DD
RF-004	Fatigue	Engagement drop > 30% WoW	Medium	List health risk	Campaign logs	Pause & refresh creative	Owner	5	YYYY-MM-DD

KPI & Formula Reference

For consistency across teams.

Formulas

Incremental lift (%) = (Treatment conv – Control conv) / Control conv × 100
Cost per incremental unit (CPIU) = Promo spend / Incremental units
ROI = Incremental revenue / Promo spend
Compliance score (%) = Average of placement, count, materials × 100

Thresholds (edit to your policy)

Lift: Green ≥ 5%, Yellow 2–5%, Red < 2%
CPIU: Green ≤ $3.00, Yellow $3.01–$5.00, Red > $5.00
Compliance: Green ≥ 90%, Yellow 80–89%, Red < 80%

Glossary of Key Terms

A

AI Velocity:

The speed at which an organization can progress from AI model development to production deployment and realize business value from artificial intelligence initiatives. Measures the time between identifying an AI use case and having a trained, tested model actively generating predictions, automations, or insights in live business operations. High AI velocity organizations can move experimental models into production within weeks, rather than months, rapidly iterating on model performance based on real-world feedback. Factors that enable AI velocity include MLOps infrastructure, standardized deployment pipelines, robust data governance frameworks, pre-validated model architectures, cross-functional AI teams with deployment authority, and executive commitment to rapid experimentation. Distinguished from general digital transformation speed by focusing specifically on the machine learning lifecycle, from data preparation and model training through validation, deployment, and monitoring. Critical for competitive advantage as AI capabilities become central to customer experience, operational efficiency, and decision making, where organizations with higher AI velocity can capitalize on emerging opportunities, respond to market shifts, and compound their AI capabilities faster than competitors still trapped in lengthy proof-of-concept cycles.

Attribution:

The process of connecting specific marketing activities to actual customer purchases or behaviors. Instead of guessing which campaigns drive sales, attribution uses data to prove which touchpoints influenced buying decisions.

Attribution Modeling:

An analytical framework that assigns credit for conversions and business outcomes to specific digital touchpoints throughout the customer journey. Determines which marketing channels, campaigns, messages, or interactions contributed to a customer's decision to purchase or take desired actions. Models range from simple last-click attribution to complex multi-touch algorithms that weight each interaction based on timing, sequence, and influence. Critical for accurate ROI calculation and budget optimization, especially when customers interact with brands across multiple digital and offline channels before converting.

C

Click-through Rate (CTR):

A digital marketing metric that measures the percentage of people who click on a specific link, advertisement, or call-to-action after viewing it. Calculated as total clicks divided by total impressions, then multiplied by 100. While CTR indicates engagement levels, it's considered a vanity metric when used in isolation because it doesn't account for incrementality or business outcomes. High CTR doesn't guarantee sales, conversions, or profitable customer acquisition. Effective measurement requires comparing CTR lift against holdout groups and connecting clicks to verified business results like purchases or qualified leads.

Collection moments:

Strategically designed customer touchpoints where brands capture first-party data in exchange for immediate value or benefits. These moments transform routine interactions (purchases, support calls, website visits, promotional events) into structured data-gathering opportunities that feel natural and rewarding to customers. Effective collection moments follow value exchange principles, providing clear benefits (discounts, personalized content, exclusive access) that motivate voluntary data sharing. Examples include QR code scans for instant rebates, receipt uploads for loyalty points, post-purchase surveys for future discounts, and progressive profiling through email interactions. Distinguished from passive data collection by explicit customer consent, transparent value proposition, and immediate benefit delivery. Essential for building comprehensive first-party customer profiles while maintaining positive brand relationships and regulatory compliance.

Consent Relationship:

A transparent agreement between a customer and organization where the customer knowingly provides personal information in exchange for clearly stated value or benefits. Requires explicit communication about what data is collected, how it will be used, and what the customer receives in return (such as personalized offers, product improvements, or enhanced service). Distinguished from passive data collection by active customer participation and clear understanding of the data exchange. Must comply with privacy regulations including CCPA, CPRA, and applicable state laws requiring opt-in consent, data use transparency, and easy withdrawal mechanisms. Forms the legal and ethical foundation for first-party data collection, ensuring customers maintain control over their information while enabling organizations to build legitimate customer intelligence. Essential for sustainable data strategies that build trust rather than exploit information asymmetries.

Correlation Conflation:

A data interpretation error where two metrics that move together are incorrectly assumed to have a cause-and-effect relationship. This fallacy occurs when analysts observe simultaneous changes in variables and conclude that one directly causes the other, without considering alternative explanations such as seasonal factors, external market conditions, or shared underlying drivers. Common in marketing analytics when campaigns launch during periods of natural demand fluctuation. Prevention requires controlled testing with holdout groups, consideration of alternative hypotheses, and statistical techniques that can isolate true causal relationships from coincidental correlations.

Cross-validation:

The process of confirming digital engagement metrics against verified purchase data to ensure measurement accuracy and eliminate false positives. Involves matching online activities (clicks, views, social interactions) with actual customer behaviors (purchases, redemptions, verified actions) to validate which digital touchpoints truly influence business outcomes. Essential for distinguishing between vanity metrics that indicate attention and performance metrics that predict revenue. Requires integrated data systems that can link digital identifiers to first-party customer records while maintaining privacy compliance.

D

Decision Lag Cost (DLC):

The financial penalty incurred when reporting delays cause continued investment in underperforming marketing tactics. Calculated by multiplying average days to verified results by daily media spend at risk and percentage of likely misallocated budget. Benchmark of 7 days or less recommended for high-velocity digital channels to minimize exposure to poor allocation decisions.

Decision-grade data:

Information that meets quality, accuracy, and reliability standards sufficient for driving business decisions and strategic actions. Characterized by verified sources, current relevance, statistical significance, and clear methodology. Must be free from significant bias, sampling errors, or data quality issues that could lead to incorrect conclusions. Typically includes audit trails, confidence intervals, and validation against known outcomes. Distinguished from exploratory or directional data by its suitability for budget allocation, product development, customer targeting, and other high-stakes business choices. Requires governance protocols to ensure consistency, compliance with privacy regulations, and cross-functional accessibility. Forms the foundation for data-driven organizations where insights directly influence resource allocation and strategic direction.

Designated Market Area (DMA):

A Nielsen-defined US media market made up of counties where most households receive the same TV stations. In planning, DMAs are used to group audiences and buy media at a regional scale that aligns with how people actually consume broadcast and many streaming ads. Example: "New York DMA," "Dallas-Fort

Worth DMA." DMAs do not equal states or ZIP codes; they often cross state lines and bundle many ZIPs into one market.

E

Execution Velocity:

The speed at which approved strategic decisions are implemented and deployed across customer touchpoints and operational systems. Measures the time between finalizing action plans and having those changes live in customer-facing channels, automated workflows, or business processes. High execution velocity organizations can deploy new campaigns, update product features, or modify customer experiences within hours or days of decision approval. Factors that enable execution velocity include automated deployment systems, streamlined approval processes, pre-built campaign templates, and cross-functional teams with clear implementation authority. Distinguished from learning velocity by focusing on implementation speed rather than insight generation speed. Essential for competitive advantage in fast-moving markets where delayed execution allows competitors to capture opportunities first.

Extract, Transform, Load (ETL):

A fundamental data integration process that systematically moves information from multiple source systems into a centralized data warehouse or analytics platform. Extract phase retrieves data from various sources (websites, CRM systems, point-of-sale terminals, survey platforms). Transform phase cleanses, standardizes, and restructures the data to ensure consistency, quality, and compatibility (removing duplicates, standardizing date formats, applying business rules). Load phase imports the processed data into the target database or data warehouse where it can be analyzed and activated. In first-party data operations, ETL processes are essential for unifying customer touchpoints into comprehensive profiles, enabling real-time identity resolution and segmentation. Modern ETL systems often operate in near real-time, processing data streams continuously rather than in batch cycles, supporting immediate activation and personalization based on fresh customer signals.

F

First-Party Data:

Information collected directly from your customers through their interactions with your brand, including website visits, purchases, survey responses, and event participation. Unlike third-party data purchased from outside sources, first-party data comes straight from your audience with their permission.

Flywheel Momentum:

The self-reinforcing acceleration that occurs in data-driven systems when each customer insight generates more effective actions, which in turn create richer customer interactions that yield higher-quality data for subsequent analysis. This compounding effect means each cycle of the insights flywheel (collect, connect, analyze, activate, measure, re-engage) becomes faster, more accurate, and more profitable than the previous cycle. Distinguished from linear processes by its exponential improvement trajectory, where early investments in data quality and systematic collection create dramatic efficiency gains over time. Essential for sustainable competitive advantage, as organizations with established flywheel momentum can adapt to market changes faster and with greater precision than competitors starting from static data foundations.

H

Holdout Group:

A randomly selected control group that is excluded from receiving a specific marketing treatment or campaign, used to establish a baseline for measuring incrementality. By comparing outcomes between the exposed group (who received the marketing) and the holdout group (who didn't), organizations can calculate the true lift generated by their marketing activities. Holdout groups are essential for proving causation rather than correlation and are considered the gold standard for marketing measurement in digital advertising, email campaigns, and promotional testing.

I

Incrementality:

A measurement principle that isolates the true business impact of marketing activities by determining what additional outcomes occurred beyond the baseline. Incrementality answers the question "What sales/conversions/behaviors happened because of this campaign that wouldn't have occurred otherwise?" Measured through controlled experiments, geo-tests, or holdout groups that establish counterfactual baselines. Essential for accurate ROI calculation and budget optimization, as it separates correlation from causation in marketing attribution.

Incremental Lift:

The additional business outcome directly attributable to a specific marketing intervention, measured as the difference between results with the intervention versus without it. Quantifies what would not have occurred naturally through baseline trends, organic behavior, or other marketing activities. Calculated by comparing a test group exposed to the marketing action against a holdout control group that receives no exposure, isolating the true causal impact. For Example, if a promotional campaign generates 10,000 conversions but the control group shows 7,000 conversions would have occurred anyway, the incremental lift is 3,000 conversions. Applied across channels, including paid advertising, email campaigns, price promotions, and personalization strategies to determine true ROI rather than correlation-based attribution. Distinguished from total lift or observed results by accounting for what would have happened without the intervention. Essential for optimizing marketing spend by identifying which tactics genuinely drive new customer behavior versus those that reach customers who would have converted regardless, preventing wasted investment on non-incremental activities.

Inventory Misallocation Cost (IMC):

The profit impact of suboptimal regional inventory distribution caused by insufficient real-time demand signals. Includes costs from both overstock situations and out-of-stock losses that could be prevented with better customer demand data. Calculated as the difference between overstock and avoided stockout units multiplied by unit margin.

L

Learning Debt (LD):

The cumulative value of optimization opportunities missed due to lack of verification systems in testing and campaigns. Represents the compound effect of not banking proven improvements because results couldn't be confirmed. Calculated as number of unverified tests multiplied by estimated lift and campaign scale. High learning debt indicates systematic underperformance relative to potential.

Learning velocity:

The speed at which customer insights translate into improved business decisions and measurable outcomes within an organization. Measured by the time elapsed between collecting first-party data signals and implementing actionable changes in marketing campaigns, product development, pricing strategies, or customer experience improvements. High learning velocity organizations can move from customer signal detection to strategic response in days or weeks, while low learning velocity organizations require months or quarters for the same cycle. Critical success factors include automated data processing, cross-functional collaboration protocols, decision-making authority at appropriate levels, and systematic feedback loops that capture results and inform subsequent iterations. Organizations with superior learning velocity gain sustainable competitive advantages by adapting to customer needs and market changes faster than competitors with slower insight-to-action cycles.

M

MPP-only opens:

Apple's Mail Privacy Protection, which artificially inflates open rates by pre-loading email content

O

Omnichannel Myopia:

A measurement bias where organizations focus heavily on digital analytics while overlooking critical offline customer touchpoints and execution factors. This creates incomplete understanding of customer behavior and campaign performance, particularly problematic for retail brands where the majority of purchases still occur in physical stores. Common symptoms include celebrating online engagement metrics while missing out-of-stock situations, poor in-store displays, or inadequate staff training that prevent digital interest from converting to sales. Prevented by integrating store-level data, point-of-sale information, and offline event feedback with digital dashboards to capture complete customer journeys across all channels.

Overall Evaluation Criterion (OEC):

A primary success metric used in experimentation that correlates with long-term business objectives. Unlike vanity metrics that measure engagement, an OEC focuses on outcomes like customer lifetime value, retention rates, or verified revenue impact. Common OECs include qualified conversions, repeat purchase rates, and incremental profit per customer.

P

Privacy Velocity:

The speed at which an organization can adapt its data practices, technology infrastructure, and customer experiences to evolving privacy regulations, consumer expectations, and consent requirements while maintaining business functionality. Measures the time between identifying a new privacy obligation - whether regulatory mandate, platform policy change, or consumer demand – and having compliant systems operational across all data collection, processing, and activation touchpoints. High-privacy velocity organizations can implement consent management updates, data deletion workflows, or cookie-less tracking alternatives within days or weeks of requirement identification, minimizing legal exposure and revenue disruption. Factors that enable privacy velocity include privacy-by-design architecture, flexible consent management platforms, centralized data governance systems, cross functional privacy teams with implementation authority, automated compliance monitoring tools, and proactive regulatory scanning capabilities. Distinguished from general compliance speed by focusing specifically on the iterative cycle of privacy requirement identification, solution design, technical implementation, and validation across the entire data ecosystem. Critical for competitive advantage as privacy regulations proliferate globally and consumer trust becomes a differentiator, where organizations with higher privacy velocity can maintain customer relationships and data-driven capabilities while competitors face enforcement actions, platform restrictions, or customer attrition due to slow privacy adaptation.

Progressive Profiling:

A data collection strategy that gradually gathers customer information over multiple interactions rather than requesting comprehensive details in a single form or survey. This approach builds trust by starting with low-barrier requests (email, name) and progressively asking for more detailed information (preferences, demographics, behaviors) as the customer relationship develops. Each additional data request is paired with corresponding value (discounts, personalized content, exclusive access) to maintain customer willingness to share. Reduces form abandonment rates, improves data quality through voluntary participation, and creates better customer experiences by avoiding overwhelming initial requests.

Essential for building comprehensive customer profiles while maintaining privacy compliance and customer satisfaction.

R

Risk-Adjusted Incremental ROAS (RA-iROAS):

A comprehensive metric for evaluating third-party data performance that adjusts traditional incremental return on ad spend calculations for real-world risks. Formula: RA-iROAS = iROAS × Quality discount × Provenance penalty × Supply volatility penalty. Quality discount accounts for invalid traffic and verification gaps. Provenance penalty reflects compliance confidence and consent clarity. Supply volatility penalty considers the probability of segment changes or vendor disruptions. This metric provides more accurate cost-benefit analysis for external data sources by incorporating hidden risks that standard ROAS calculations ignore.

S

Sales Velocity:

The speed at which an organization converts prospects into customers and generates revenue, calculated as the product of operability volume, win rate, average deal size, and sales cycle length. Measures the rate of revenue flow through the sales pipeline, quantifying how quickly qualified leads transform into closed deals and recognized income. High sales velocity organizations generate more revenue in less time by optimizing each component- increasing lead quality and quantity, improving conversion rates at each pipeline stage, expanding average contract values, and compressing time-to-close through streamlined processes. Factors that enable sales velocity include predictive lead scoring systems, sales enablement content that accelerates buyer decisions, automated proposal generation, pre-negotiated contract templates, cross-functional deal teams that remove friction points, and real-time pipeline analytics that identify bottlenecks. Distinguished from simple sales cycle time by incorporating deal quality and volume metrics, recognizing that closing deals faster means nothing if win rates collapse or deal sizes shrink. Essential for competitive advantage in markets where buyer preferences shift rapidly and early market share establishes category leadership, where organizations with higher sales velocity can outpace competitors for customer acquisition, achieve faster returns on marketing investment, and compound growth advantages through earlier customer relationships and feedback loops.

Sample-size Distortion:

A statistical error where conclusions are drawn from test groups too small to be representative of the broader population or market conditions. Small samples are highly susceptible to random variation and outlier effects that disappear when scaled to larger audiences. Common in marketing when promising pilot results from limited geographic areas, short time periods, or narrow customer segments fail to replicate across broader rollouts. Symptoms include dramatic performance swings that normalize at scale, results driven by individual high-performers rather than systematic improvements, and statistical significance that vanishes with larger sample sizes. Prevention requires pre-setting minimum sample size thresholds, power analysis, and replication across multiple test conditions before scaling decisions.

Service Level Agreement (SLA):

A formal commitment that defines specific performance standards, response times, and quality metrics for delivering services or completing processes within an organization. In the context of first-party data operations, SLAs establish measurable targets for moving insights through the INSIGHT Framework stages, such as maximum time between data collection and analysis, required handoff documentation, and minimum accuracy thresholds for customer segmentation. SLAs create accountability by specifying consequences for missed targets and escalation procedures for resolving bottlenecks. Essential for maintaining learning velocity in data-driven organizations, as they prevent insights from stalling between departments and ensure systematic progression from customer signals to business actions. Typically includes metrics like processing time limits, data quality requirements, and availability standards that enable cross-functional teams to coordinate effectively while maintaining operational discipline.

Shapley Additive Explanations (SHAP):

A machine learning interpretability method that quantifies each input feature's contribution to individual model predictions by applying game theory principles to determine fair attribution of predictive power. Calculates how much each variable, such as customer demographics, behavioral signals, or transaction attributes, increases or decreases the predicted outcome compared to a baseline expectation, providing both magnitude and direction of impact for every feature in every prediction. Delivers consistent, theoretically grounded explanations by computing Shapley values, which represent the average marginal contribution of each feature across all possible combinations of other features, ensuring that contributions sum precisely to the difference between the prediction and baseline. Applied across classification and regression models, including tree-based algorithms, neural networks, and ensemble methods, to answer critical business questions like "which factors drive this customer's churn probability?" Enables model debugging by surfacing unexpected feature relationships, supports regulatory compliance by documenting automated decision logic, and builds stakeholder trust by making black-box predictions transparent. Distinguished from simpler feature importance methods by providing prediction-level explanations rather than only global model summaries, and from other local explanations techniques by guaranteeing mathematical properties like consistency and local accuracy. Essential for deploying AI in regulated industries, high-stakes decisions, and customer-facing applications where explainability requirements, fairness audits, or user trust demand understanding not just what the model predicts but precisely why.

Stage Handoffs:

The formal transfer of deliverables, accountability, and decision-making authority between consecutive stages of the INSIGHT Framework, characterized by defined timing requirements, quality criteria, and documentation standards. Each handoff includes specific artifacts (data models, segment definitions, insight briefs, activation plans), measurable completion criteria, and designated receiving parties who accept responsibility for the next stage. Effective stage handoffs prevent bottlenecks by establishing clear expectations for when work products are considered complete and ready for progression. Include escalation protocols for resolving quality issues or timing delays that could disrupt the insights-to-action cycle. Essential for maintaining learning velocity in data-driven organizations, as poor handoffs create delays, duplicated work, and accountability gaps that slow competitive response times. Typically governed by Service Level Agreements (SLAs) that specify maximum transition times and minimum deliverable standards.

T

Third-Party Data:

Customer information purchased from external vendors, data brokers, or platforms rather than collected directly from a company's own audience interactions. Includes demographic profiles, behavioral segments, lookalike audiences, and intent signals sourced from websites, apps, and services outside the purchasing organization. While third-party data offers apparent scale and reach, it introduces risks including compliance exposure, segment inaccuracy, supply volatility, and competitive parity. Contrasts with first-party data, which companies collect directly from their customers with explicit consent and full control over methodology and quality.

U

Unverified Spend Rate (USR):

A financial metric that measures the portion of marketing budget allocated to promotional activities that cannot be directly linked to customer purchases or verified business outcomes. Calculated as unverified promotional spend divided by total promotional spend. High USR indicates significant budget waste and missed learning opportunities. Organizations typically aim to reduce USR below 15% through systematic verification mechanisms.

V

Verification:

Confirming that promotional activities generated real customer outcomes through measurable proof points like receipts, purchase data, or tracked behaviors. Verification turns assumptions into facts by linking marketing spend to business results.

Bibliography

1. 1WorldSync. Consumer Survey Data: The Growing Impact of QR Codes and AI on In-Store Shopping Experiences. Dec 10, 2024. 1WorldSync

2. Attorney General of Texas. Texas Data Pri va cy And Secu ri ty A ct. https://www.texasattorneygeneral.gov/consumer-protection/file-co nsumer-complaint/consumer-privacy-rights/texas-data-privacy-and-sec urity-act

3. AVAU. First-party data benchmarks – what results can you expect? https://www.avaus.com/blog/first-party-data-benchmarks/

4. Bitly. From Scans to Strategy: How Marketers Use QR Codes. n.d. Accessed Aug 22, 2025. Bitly

5. Bloomberg Law. Which States Have Consumer Data Privacy Laws? April 7,2025. https://pro.bloomberglaw.com/insights/privacy/state-p rivacy-legislation-tracker

6. Colorado Attorney General. Universal Opt-Out Mechanism (UOOM). Official page and recognized list. Colorado Attorney General

7. Colorado Office of the Attorney General. Universal Opt-Out Mechanism (UOOM). n.d. Accessed Aug 22, 2025. State of Colorado, Department of Law. https://coag.gov/uoom/

8. CookieYes, Third-Party Cookies Going Away? Here's What's Actually Happening, May 28, 2025. https://www.cookieyes.com/blog/third-party-cookies-going-away

9. Compliance windows: ADMT (1/1/2027), risk assessment deadlines (12/31/2027; attestations 4/1/2028), cybersecurity audit cadence. https://perkinscoie.com/insights/blog/cppa-approves-cybersecurity-automated-decisionmaking-and-risk-assessment-0https://www.onetrust.com/blog/cppa-adopts-new-regulations-what-businesses-need-to-knowhttps://www.bakerbotts.com/thought-leadership/publications/2025/august/a-101-of-the-cppas-finalizes-rules-on-admt-risk-assessments-and-cybersecurity-audits

10. CPPA Board approval of ADMT/Risk/Cyber regs (July 24, 2025) & rule package timeline. chrome-extension://efaidnbmnnnibpcajpcglclefindmkaj/https://cppa.ca.gov/meetings/materials/20250724_item5_fsr_draft.pdf

11. EXP Platform. Pitfalls of Long-Term Online Controlled Experiments. https://exp-platform.com/pitfalls-of-long-term/

12. Google Privacy Sandbox team. Next steps for Privacy Sandbox and tracking protections in Chrome, Apr 22, 2025. plus developer docs on the Chrome grace period and tracking protections. Privacy Sandbox

13. GS1. Sunrise 2027, https://www.gs1us.org/industries-and-insights/by-topic/sunrise-2027

14. IAPP25 Resource Center. US State Privacy Legislation Tracker. Jul 7, 2025. https://iapp.org/resources/article/us-state-privacy-legislation-tracker/

15. IAPP News. CPPA board finalizes long-awaited ADMT, risk assessment rules, Jul 25, 2025; Morgan Lewis client update, Aug 4, 2025. IAPP-

Morgan Lewis

16. informs PubsOnLine. Frontiers: How Effective Is Third-Party Consumer Profiling? Evidence from Field Studies, Oct 2, 2019 https://pubsonline.informs.org/doi/abs/10.1287/mksc.2019.1188

17. International Association of Privacy Professionals. "US State Privacy Legislation Tracker." Updated July 7, 2025. IAPP

18. McKinsey & Company. The value of getting personalization right, or wrong, is multiplying. Nov 12, 2021 https://www.mckinsey.com/capabilities/growth-marketing-and-sales/our-insights/the-value-of-getting-personalization-right-or-wrong-is-multiplying

19. **ModelOps**: backed CI/CD/CT, drift monitoring, and pre-deployment testing with Google's MLOps guidance, TFX, and AWS Model Monitor/Clarify. https://cloud.google.com/architecture/mlops-continuous-delivery-and-automation-pipelines-in-machine-learning

20. PGM. The Power of First-Party Data. Jul 27, 2025 https://porchgroupmedia.com/blog/first-party-data/

21. QRCodeChimp. First-Party Data Statistics Every Marketer Should Know Jan 2, 2025 https://www.qrcodechimp.com/first-party-data-statistics/

22. Thales Cybersecurity. 2025 Bad Bot Report, https://www.imperva.com/resources/resource-library/reports/2025-bad-bot-report

23. UOOM/GPC recognition & enforcement: Colorado (GPC recognized; 7/1/2024 start), Connecticut (1/1/2025), New Jersey (7/15/2025).
https://www.foley.com/p/102iw4p/colorado-state-ag-approves-global

-privacy-signal-as-first-universal-opt-out-mechttps://portal.ct.gov/ag/press-releases/2024-press-releases/tong-advises-connecticut-consumers-and-businesses-of-opt-out-rights-and-requirementshttps://www.njconsumeraffairs.gov/ocp/Pages/NJ-Data-Privacy-Law-FAQ.aspx

24. U.S. Census Bureau, Center for Behavioral Science Methods. "Usability of Quick Response (QR) Code as a Method to Access a Survey Using a Smartphone (Survey Methodology #2024-05)." Aug 2024. U.S. Census Bureau Working Paper. Census

Index

About the Author

Lisa L. Fagen is the Founder and CEO of Promolytics, a SaaS platform that helps brands capture real-time, first-party consumer insights and turn them into measurable growth. Promolytics links QR scans, receipt uploads, surveys, and link clicks to a privacy-compliant CDP, building unique consumer profiles. Each touchpoint is traced to its source, allowing marketing teams to precisely segment, build data-driven cohorts, and repeatedly retarget with effective messages, thereby fueling the campaign flywheel.

With over two decades of experience in sales, marketing, data analytics, strategy, and e-commerce, Lisa has built her career at the intersection of business growth and customer understanding. She has held roles at industry leaders, including Republic National Distributing Company (RNDC) (the second-largest wine and spirits distributor in the United States), an Anheuser-Busch distributor, and Health Services Corporation of America (HSCA). She is recognized for driving revenue growth, advancing data-driven marketing and analytics, and connecting traditional and digital channels. Her work has delivered measurable results, from increasing customer engagement to turning around major accounts and leading cross-functional teams.

Her expertise in translating strategy into practical commercial execution has earned her multiple honors, including three Sales Representative of the Year awards and being named Top Sales Rep in her class, establishing her as a trusted partner to brands seeking market-driven, first-party insights.

Lisa L. Fagen: https://www.linkedin.com/in/lisa-fagen-promolytics/

www.ingramcontent.com/pod-product-compliance
Lightning Source LLC
Chambersburg PA
CBHW071320210326
41597CB00015B/1294